U0303668

[加拿大]詹妮弗·内格尔 著　徐竹 译

知识

牛津通识读本·

Knowledge

A Very Short Introduction

译林出版社

图书在版编目（CIP）数据

知识 ／(加) 詹妮弗·内格尔 (Jennifer Nagel) 著；
徐竹译. —南京：译林出版社，2022.1
(牛津通识读本)
书名原文：Knowledge: A Very Short Introduction
ISBN 978-7-5447-8883-0

Ⅰ.①知… Ⅱ.①詹… ②徐… Ⅲ.①知识论 Ⅳ.
①G302

中国版本图书馆 CIP 数据核字 (2021) 第 210079 号

Knowledge: A Very Short Introduction, First Edition by Jennifer Nagel
Copyright © Jennifer Nagel 2014
Knowledge was originally published in English in 2014.
This licensed edition is published by arrangement with Oxford University Press.
Yilin Press, Ltd is solely responsible for this bilingual edition from the original work and Oxford University Press shall have no liability for any errors, omissions or inaccuracies or ambiguities in such bilingual edition or for any losses caused by reliance thereon.
Chinese and English edition copyright © 2021 by Yilin Press, Ltd

著作权合同登记号 图字：10-2017-080 号

知识 ［加拿大］詹妮弗·内格尔／著 徐 竹／译

责任编辑 许 丹
装帧设计 景秋萍
校　　对 戴小娥
责任印制 董 虎

原文出版 Oxford University Press, 2014
出版发行 译林出版社
地　　址 南京市湖南路 1 号 A 楼
邮　　箱 yilin@yilin.com
网　　址 www.yilin.com
市场热线 025-86633278
排　　版 南京展望文化发展有限公司
印　　刷 江苏扬中印刷有限公司
开　　本 635 毫米 ×889 毫米 1/16
印　　张 17.5
插　　页 4
版　　次 2022 年 1 月第 1 版
印　　次 2022 年 1 月第 1 次印刷
书　　号 ISBN 978-7-5447-8883-0
定　　价 39.00 元

版权所有·侵权必究

译林版图书若有印装错误可向出版社调换。质量热线：025-83658316

序　言

陈嘉映

从理知时代起，人们就开始反思知和知识的性质，且不说柏拉图、亚里士多德，孔子、墨子、老子、庄子、荀子也都有关于知的深刻思考。当然，直到近世，知识论才形成一个学术分支，在分析哲学学统那里，知识论又围绕某些特定的视角和问题展开。本书浅近易解、简明扼要地介绍了知识论领域里的这些思考、视角和问题，既包括经验主义、理性主义、怀疑论这些十分广泛的观念，也包括盖梯尔反例、语境主义、封闭原则这样一些相对专门的概念。短短一百多页，可以让读者对当代知识论关注的一大批主要问题有个大概了解。这本书的另一个优点是译文准确、平实。

不过，我个人对主流知识论有相当的保留，愿意借这个机会提示其中几点，也许对阅读此书的读者不无帮助。

第一，分析哲学追求清晰，这当然是个优点，然而，很多分析哲学家是以科学为榜样来追求清晰的，尝试把思想困惑转变成一些类科学问题。这个方向上的努力带来的往往是徒有其表的清晰。科学问题是有解的，其外在标志是：在大多数情况下，科学问题的提出方式和解决会达到共识。这既不是哲学思考的目标，也不是哲学思考的实际结果——即使在分析哲学内部，研究者也从来没有在哪个解答上达至共识。个中缘由牵涉太广，无法在这里展开。

第二，对“知识”的探究有种种途径，例如，对知识、知、看法、信念等的概念探究，对获取知识的心理过程的研究，以及对这类心理活动的神经活动基础的探究。我们一开始就须记得它们是不同层级的探究。我不是说这些不同层级的探究各自隔离，不能互相启发，但我的确认为我们不应该不加注明地把它们共置一炉。

第三，这本小书没有讨论自我认识这个课题。的确，主流知识论主要依循的是对象性认识的思路，自我不是一个现成摆在那里的单纯对象，我们甚至不会把自我认识的结果叫作“（关于自己的）知识”。然而，实际上知识论绕不开自我认识问题，因为对知识的反思必定涉及“关于知识的知识”或“我们对知识都知道些什么”，只不过，作者立刻把这个话题转向“读心”问题（第112页）。但对“他心”的认识不可能脱离对自己的认识。对象性认识在面对自我认识和“他心”问题时却难免左支右绌，最后只能求助于“对心理状态的表征”之类的途径。这类途径是远远不足用的，因为人的心灵，无论自我还是他心，都不可能被充分对象化。毋宁说，自我认识本身就是自我如何存在的一部分：你认识到自己的能力和缺陷，你认识不到自己的能力和缺陷，你将是不同的人。而这一类更富意趣的话题通常都逸出了主流知识论的视野之外。

目　录

致　谢　1

第一章　引　言　1

第二章　怀疑论　13

第三章　理性主义与经验主义　33

第四章　知识的分析　50

第五章　内在论与外在论　65

第六章　证言知识　78

第七章　标准转换？　94

第八章　关于知识的知识　112

索　引　128

英文原文　133

致　谢

我要感谢埃莱娜·德克森、爱玛·马、艾利克斯·泰伦鲍姆、赛吉欧·泰伦鲍姆和蒂姆·威廉姆森，给予我许多有益的评论和意见。我还要感谢加拿大的人文与社会科学研究委员会对这项工作的支持。

第一章

引　言

知识的寻求

获取知识现在可以说是再简单不过的事情，即便是很难的问题也只需要敲几下键盘就能获得答案。我们个人的记忆、知觉与推理能力可以毫不费力地由很久之前的朋友或专家来检验。如果过往的几代人依然在世，他们也一定会为我们所能获取的书籍数量之巨大而无比惊讶。

然而，这些新的优势并不能使我们避开一个古老的问题：如果知识如此容易获得，那么单纯的意见也是，而我们很难分辨哪些是真正的知识，哪些只不过是一些意见。譬如，看起来值得信任的网站也许并不客观公正，享誉世界的权威也可能受某些证据的误导而走向错误的方向，幻觉还会扭曲我们自认为看到或记得的事情。因此，那些初看起来像是知识的东西，也有可能最后被证明并非真的知识。考虑到这项探究的困难，我们不免怀疑真正的知识究竟是什么样的。知识究竟是什么？在认为某物为真与真正知道它为真之间，究竟有什么差异？我们究竟如何能获得知识？

这些问题都非常古老，而且有一个专门的哲学分支来回答它们——知识论（epistemology），几千年来一直是较为活跃的哲学主题。某些核心的问题历经千年的思考仍然亘古常新：知识与真理的关系是什么？视觉和听觉这样的感觉提供知识的方式是否与抽象推理提供知识的方式相同？为了确定知道某件事情，是否需要具备证成该项主张的能力？有的研究课题则是近来才涌现的，主要是根据某些人性、语言和心灵上的新发现。譬如，知识与意见的对立是否存在于所有文化之中？一般而言，“知道”这个词总是指称相同的对象吗？还是说它在法庭辩论中指称意义重大的事实，而在公交车站的随意交谈中指代无足轻重的东西？从自然的直觉上看，我们对于他人所知的东西究竟拥有怎样的印象？这些印象又告诉了我们多少关于知识本身的东西？

1

几个世纪以来，哲学家们在对知识的探究中发掘出了许多费解的谜题与悖论。他们同时也创造性地发展出了这些问题的解决方案。本书将首先回顾知识论历史上的这些重要进展，而后就将引向这一领域在当代的核心论争。以下我们先从回顾知识的一些特征开始，这很容易激发哲学上的好奇心。

知识与认知者

知识常被勾画为某种自由流动的客观资源：知识被认为储存在数据库与图书馆里，通过“知识经济”——有时也被称为“信息驱动的商业”——发生交换。如许多资源一样，知识可以是为了多重目的而被获取与应用，也可能会丢失——有时要付出很高的代价。但是，知识与我们之间的关系要远远比水或黄金那样的资源与我们的关系更为密切。即使人类的感性生活在某一灾难中被完全抹杀，黄金仍然可以继续存在。然而，知识是否能持续存在下去，却不能不依赖于认知者的存在。

知识还很容易被等同于事实，但并非所有事实都是知识。设想一下，晃动一只装有一枚硬币的密封纸箱，当你放下纸箱，2 其中硬币的两面之中必有一面朝上：我们把这称作一个事实。但只要还没有人打开过纸箱，这个事实就仍不为人所知；它就还不是一项知识。此外，仅仅通过写下来，事实也不会由此变为知识。假设你在一张纸上写下："这枚硬币正面朝上"，而在另一张纸上写"这枚硬币背面朝上"，那么这两张纸上所写的总有一个是事实，但你仍然不会知道究竟硬币的哪一面是朝上的。所以，知识所要求的乃从主体的一方通达事实的某种方式。只要不考虑如何通达事实，则不论图书馆或数据库里保存的是什么，它们都不会是知识，而只不过是一些墨水符号或电子印记。就任何已知的知识而言，对事实的通达有可能专属或不专属于个人：同一个事实或许只是由某个人所知道，而并不为他人所知。普通的知识或许能够被多人分享，但绝没有任何知识空荡荡地摇晃而不从属于任何主体。与水或黄金不同的是，知识永远从属于认知者。

更确切地说，知识总是从属于某些个体或群体：群体的知识或许会超越其中个体成员的知识。有时一个群体知道某个事实，只是因为该群体的所有成员都知道了这个事实（"乐队知道音乐会晚上八点钟开始"）。但有时我们也会说，乐队知道如何

图1 所知物必须关联到认知者

演奏贝多芬的《第九交响曲》全篇，即便其中的个体成员仅仅知道他们自己所负责演奏的部分。又或者说，某个流氓国家知道如何发射搭载核弹头的导弹，即便这个国家中没有任何一个人知道控制这一发射所需要的哪怕一半信息。群体可以以极具创造性（或破坏性）的方式组合其成员的知识。

那么，是否存在超越人类个体与群体的知识？我们是否有必要讨论非人类的动物的知识？或者，假如有上帝的话，还有上帝的知识？这些问题极有可能把我们推向生物学和神学上的艰难论争。出于这个原因，大多数知识论学者都从简单的范例出发，即某个个人（譬如你自己）的知识。这种知识将会是本书关注的焦点。这里所关心的意义上的知识，即个人与事实之间的关联。即便我们把关注点限制到某一个认知者与某一个所知的事实上，描述这种关联仍然非常有挑战性。对你来说，知道某件事情，而非仅仅相信它，是什么意思？

发现差异

如果要寻求真正的知识与一般的信念之间的差异，首先需要考虑的是，我们如何能确信这里真的**有**某种差异。考虑一下这样的观点：知识与意见之间没什么真正的区别。倘若“知识”不过是我们贴在精英态度上的标签呢？在我们的文化中，或许诺贝尔奖得主的科学研究就是这种意义上的知识，或是某个首席执行官对他的企业的思考也是知识；而在历史上的其他时代或世界上其他地域，大祭司的教义或部落长者的意见也可以是这种意义上的知识。与此相一致的是，弱者或处于劣势的人们持有的观点却被看作迷信或误解。按照这种犬儒主义的理论，某个观点究竟是知识还是仅仅作为意见，取决于持有观点的人究竟是领袖精英还是竞争的失败者，而无关乎这一观点本身及其与实在的关系。

把知识看作地位的标签，也并非全无道理。“知识”肯定是一个有吸引力的标签，说某种态度是知识，就意味着它高于许多其他态度。而且，知识与权力之间也的确存在着很强的双向关联：权力通常会使人具备一些优势，有助于获取知识；而知识也能帮助人们攫取权力。我们对知识的判断甚至也常常会受到某些偏见的影响，这些偏见通常也正是来自所评价对象的社会地位。但犬儒主义理论所主张的要远远超过这些，即权力与知识总是携手并进，或者说人们通常认为两者密不可分——犬儒主义主张知识只不过是对权力的感知，除此之外，别无他物。

犬儒主义理论试图解释“知识”一词的实际使用，但它没有把握住某些相关事实。首先，它低估了人们对抗有权者观点的能力：即便是诺贝尔奖得主的观点也不能免于怀疑和挑战。或者更一般地说，即便我们很有可能会视有权者为有知者，我们也依然会在真正的知识与仅仅看似有知识之间识别出区分的界限。其实有很多这样的例子，其中曾经被认为是有知识的专家，结果却被证明是错了。其次，它也没有抓住我们在日常生活中谈论知识的方式。“知道”这个动词并不只是顶尖的专家或领袖人物才能用得上，它在英语中是十个最常用的动词之一。通常我们描述看到的、听到的和回忆起来的事情，都是默认用这个动词。例如，你知道昨天晚上吃了什么，也知道谁赢得了最近一次的美国总统大选，更知道你现在是不是穿着鞋子。

在这方面，英语并不是特例。在诸如俄语、汉语、威尔士语和西班牙语等许多语言中，“知道”这个词都是最常用的动词之一。而英语与很多其他语言一样，还有一个容易引起误解的特征：在英语中，“know”这个词有两个不同的含义。一方面，它后面可以跟命题内容或“that”从句（例如，“他知道[that]那辆车失窃了”），或是某个嵌入的问题（例如，“她知道谁偷了那辆车”，或“她知道偷窃行为是什么时候发生的”）。另一方面，“know”后面也可以是直接宾语（“他认识巴拉克·奥巴马”；“她知道伦敦”）。有的语言则以不同的词表达这两个含义，例如法语中的“savoir”和“connaître”。下面，我们主要集中于考察“知道”的第一个含义，也就是那种把人与事实联系起来的知识。

有意思的是，这种意义上的“知道”在世界上所有的6 000多种人类语言中都有对应的词（“认为”也有着相同的地位）。这其实是非常罕见的。一个受过良好教育的人大概能掌握20 000个单词，但其中大概只有不到100个能够在所有其他语言中有准确的翻译。有些非常常用的词汇，譬如“吃”和“喝”，你或许认为在所有语言中都找得到，但其实并不总是能找到意义相同的对应词。（某些澳大利亚和巴布亚新几内亚的土著语言就不区分“吃”与“喝”，它们只有同一个词，意思是“咽下”。）有时却又反过来，其他语言做出的是更细致而非更粗略的区分。例如，很多语言中并没有单一的词汇对译“走”，因为它们用不同的动词分别指称行走这样的自主行动与借助交通工具的移动。而在有些情况下，区分的边界却又划在不同的地方：例如，英语中普遍用代词“他”和“她”来区分性别，而在其他语言中，第三人称代词区分在场与不在场的人，而不在性别上做区分。人类的语言是极其丰富多样的。尽管如此，某一些词汇仍然出现在所有已知的语言之中，或许是因为它们的含义对语言的作用来说非常关键，又或许是因为它们表达了人类经验中某些关键的方面。这些共同的词汇就包括“因为”“如果”“好”“坏”“生”“死”等等，以及“知道”。（见方框1）

6

方框1　在所有语言中共同与非共同的词汇

共同词汇	非共同词汇
知道，认为，看，想要，听，说	吃，喝，停止，打，走，坐
我，你	她，他，他们
好，坏	幸福，悲伤
不，可能，因为，如果，真实，前，后	栽种，树，动物，鸟，冷，热

知道与认为

那么，对于如此重要的动词，我们通常又是怎样使用的呢？“知道”究竟如何区别于“认为”？语词的日常使用会提供某些线索。考虑下面这两个句子：

吉尔知道她锁上了门。

比尔认为他锁上了门。

我们立刻就能注意到，在吉尔与比尔之间有某种区别。那7么这里的差异究竟是什么呢？我们所能想到的一个因素是与“锁门”那个内嵌命题的真值有关。如果比尔只是认为门锁上了，也许是因为门并未真正上锁，譬如说他在早晨离开家的时候没有让钥匙在锁孔里多转一圈。但吉尔的门却必须是锁上了的，因为她所说的那个命题必须为真：通常当我们说“吉尔知道她锁上了门”时，我们一般不能接着说“但她的门没上锁”。正是知识把主体与真值联系起来。“知道”的这个特征被称为**事实**

性（factivity）：我们所能够真正知道的只有事实或真命题。“知道”并不是唯一有事实性的语词用法，类似的还有“意识到”、“看到”、“想起来”以及“证明出来”等。例如，只有当你真的中了彩票的时候，你才可能意识到你中了彩票。因此，“知道”的特殊性之一，正在于它体现了这一类动词的共性，代表了像“想起来”“意识到”等动词共同具有的深层状态。显而易见，看到谷仓起火或是证明出不存在最大的质数，都不过是两种获得知识的具体途径。

当然，有可能我们只是**好像**知道某些事情，而后却被证伪了——但只要我们认识到其中的命题为假，我们就不会再主张自己先前知道。（“我们原以为他知道，但最后证明他其实是错了，根本不知道。”）更为复杂的是，我们似乎很难区分真正拥有知识与只是看似知道的情形，但这并不意味着能够抹杀两者之间的区别。在一个充斥着假货的市场里，人们也会难以分辨真钻石与假钻石，但这只是一种实践上的困难，并不是说其中不存在任何区别：因为毕竟真钻石所具备的特殊本质——碳原子的特殊结构——乃是假钻石所不具备的。

知识的本质之一就在于这种与真值的特殊联系。我们当然有时也会说“知道”某些假命题，但此时我们恰恰不是在字面意义上使用这个词。正如我们可以反讽地使用“美味”这个词，来描述某些实际上非常难吃的食物。强调（斜体或加重语气）往往标志着这种非字面意思的用法。例如，“那道卷心菜汤闻起来很**美味**，是吧？”“我**知道**自己入选了参赛队伍，但最后发现原来我并没有被选中。”这里的“知道”就是投射性的用法：说话者乃是把自己投射到一个过去的心境中，回忆起某个时刻，那时他自以为已经知道。语气上的加重表明，说话者已经把他自己区别于那个他所投射于其中的心境：在过去的那个时刻，他并没有真正地知道，从而他也不是在字面意思上使用“知道”。卷心菜汤的例子也是同样的，说话者加重语气正是表明他其实并不喜欢这道汤。所以，如果使用“知道”的字面意思，那么它就不会以上述方式关联虚假的命题。

相反地，主体却很容易通过信念关联虚假的命题。我们完全可以这样说：“比尔认为他锁了门，实际上并没有。”“认为”这个动词是**非事实性**的。其他非事实性的动词还有“希望”、“猜想”、“怀疑”以及“说”——当门没上锁的时候，你完全可以确定地说它锁上了。非事实性的看法并不见得总是错的。如果比尔认为他锁上了门，他也有可能是对的。或许他有个不靠谱的室友鲍勃，有时会忘记锁门。如果比尔不能完全肯定门上锁了，他就只能是认为它锁上了，而且也可能是正确的，只不过他无法知道门上锁了。信心对知识而言非常重要。

除了真值与信心以外，知识还有其他的要求。假设有的人非常确信某件事情，却是基于某些错误的理由，那么他就依然无法知道。譬如，一位父亲的女儿被指控犯了罪，他可能完全确定地相信女儿的无辜。但如果他的这种确信乃是基于情感而非证据（假设他故意不看任何与此案相关的事实信息），那么即便他是对的，他的女儿最后被证明是清白的，这位父亲也可能并不真的知道女儿是无辜的。但假若确信某个真信念仍然不足以使其成为知识，那么知识还需要添加什么东西？这已经被证明是个非常难解答的问题，足以需要整整一章（第四章）来专门处理它。

命题之真既然对知识的本质如此重要，这里免不了要多说

9

几句。我们通常都假定命题之真是客观的，或者说以实在为基础，且对所有人都是一致的。大多数哲学家都同意真值的客观性，但也有一些持不同意见的反叛者。古希腊哲学家普罗塔哥拉（公元前5世纪）主张，知识虽然总是真的，但不同的东西也可以相对于不同的人而为真。例如，在一个凉风习习的夏日站在屋外，且又感到有些许不适，我就知道吹来的是冷风，而你却知道吹来的是暖风。当然，我知道风对我来说是冷的，而你知道它对你来说是暖的——这其实就是说不同的人感受会有差异，这对于主流观点而言完全不是问题。赞同有这些差异，并不会妨碍我们主张命题之真对所有人都是一致的。（即便是暖风也可以让身体不舒服的人感到寒凉，这几乎是一个非常明显的客观事实。）但普罗塔哥拉所主张的并不仅限于此，而是更为激进：对我而言，“吹来的实际上**是**冷风”为真，对你而言“吹来的**是**暖风”也为真。实际上，普罗塔哥拉从来都把命题之真理解为相对于主体而言的：某些事情对你为真，另一些事情则是对你的好朋友为真，或是对你的死敌为真，而没有那种不针对任何主体、简单地为**真**的事情。

普罗塔哥拉的相对主义知识论很有趣，但也很难让人信服，甚至它本身就是自我否定的。如果事情本身对每个人来说都是自己看到的样子，那么也就无人会犯错。若按此推理，沙漠中的旅客在出现幻觉时看到的绿洲，对他而言就是真实存在的；而对把7加5错算为11的人来说，“7+5=11”也就为真。况且，如果随后你发现自己犯了错误，又该如何呢？如果事情总是它们所显现的样子，那么对你来说你真的犯了个错误，纵然显相绝不会误导人，因此一开始你也不可能犯错。这显然是非常别扭的处境。

古希腊哲学中处理这个问题的策略是区分“此刻的你”与“此刻之前的你”。有的事情只对当下的你为真，而另一些事情则可能对今后的你为真。例如，你过去犯了个错误，这可能就对未来的你为真。

让自我分裂为时刻上的片段，这对知识论而言恐怕是个高昂的代价。如果你觉得这个代价难以承受，还想保持一个持续性自我的观念，那么普罗塔哥拉的理论可能就需要放弃。如果普罗塔哥拉的理论看似为假，那么它就真的有麻烦了——不要忘了，正是这一理论告诉我们，我们不会犯错，因为事情总会是它们展现给我们的样子。这恰好表明，相对主义的观点具有自我解构的潜在可能性。柏拉图（约公元前428—前348）也已经指出，在普罗塔哥拉的理论试图要做的事情，与其主张命题之真总是相对于个人的观点之间，存在着显著的张力。一方面，普罗塔哥拉主张命题之真总是相对于每一时刻的每一个人；另一方面，他又试图用这一理论概括对所有人都成立的真理，这本身就让人无从知晓他何以能做到了。

自普罗塔哥拉以降，相对真理的概念还有很多深思熟虑的捍卫者，但大多数哲学家还是更倾向于客观真理的概念。为真的东西总是对我们所有人都为真，仅此而已，无论我们是否意识到了这一点。如果我们一定要从视角来谈这一点，那么为真的东西就是在上帝视角中呈现的事物。但客观的真理是人类所能认识的吗？怀疑论者对这一点提出了诸多质疑。

第二章

怀疑论

你能肯定吗？

设想一件你所知道的最不起眼也最容易验证的事实。譬如，你知道你现在是否穿着鞋，对吧？

然而，怀疑论者会请你三思。有没有可能你只是梦见自己正在读这本书？如果这只是一场梦，你也许正光着脚躺在床上；又或者是你穿戴整齐却在通勤的列车上睡着了。你也许会觉得自己现在不可能是在做梦，但你或许会想知道，如何能确定无疑地证明自己现在是清醒的呢？如何确定所有事情正是它们所展现的那样呢？也许你记得在什么地方读到过，只要掐自己一下就能从梦中醒来，但你所读到的材料是否值得信赖？或者它究竟是不是你所**真正**读到的东西，区别于你只是现在梦见自己曾经读到的东西？假如你无法确定地证明自己确实是醒着的，那你是否能“如其所是”的那样接纳自己的感觉经验呢？

一旦你开始对自己所知的东西有自我意识，即便是最简单

图2　你知道什么?

最不起眼的所知事实，通常你觉得一眼就能检验出来的东西，你也会开始感到其实你并不真正知道它。那些你曾想当然地认为是证据的东西，突然变得很可疑，你会开始觉得确定性完全变得遥不可及。几个世纪以来，这种思维模式让许多哲学家都困扰不已，足以让人们深深地怀疑，究竟人类是否能有真正的知识。这些哲学家就是“怀疑论者”（sceptics），这个词来自古希腊语的“探究”或“反思性”。

怀疑论的历史根源

事实上，古希腊哲学就产生了两个怀疑论传统：学院派怀疑论和皮罗主义怀疑论。学院派的怀疑论者想要证明的结论是，知识本身是不可能的；皮罗主义怀疑论者则根本不想得出任何结论，而是要在所有问题上都悬搁判断，甚至包括知识的可能性问题。

学院派怀疑论取名自它诞生的研究机构：由柏拉图创立的雅典学园。这一运动的两大领袖同时也在领导着学园：一位是公元前3世纪的阿凯西劳斯，另一位则是百年之后的卡尼底斯。虽然他们的怀疑论思想与当时影响甚大的斯多葛学派知识论的观点相左，但直到今天，他们的怀疑论论证仍然在哲学上被认真地对待。他们的怀疑论论证之所以仍有效力，是因为他们所批评的斯多葛学派核心观点，仍然是许多其他知识论立场的题中之义，甚至也存在于我们对知识与信念区别的常识思考之中。

13

斯多葛学派的知识论主张区分印象与判断。这一派学者注意到，人们可以有印象而无判断。譬如说，我们看到沙漠中有水闪烁着微光，却并不断定实际的情况正如它显现的样子。判断是对某个印象的接受或拒斥；知识是明智的判断，或者说是对正确印象的接受。在斯多葛学派的观点中，当人们因为犯了某些错误而接受了错误的印象时，他们就没有知识。譬如，如果你断定有某个朋友正向你走来，而这只是基于你远远地、模模糊糊地瞥见有人正在走来，那么你所得到的印象就不清晰，你就有可能犯了错误。或者说，即便你是对的，你也只是碰运气蒙对的，而运气好蒙对了的东西并不能算作知识。明智的人会等到那位朋友再走近些，看清楚之后再做判断。所以，按照斯多葛学派的观点，要想获得知识，你所接受的那个印象必须足够清晰和准确，以至于你根本不可能出错才行。

学院派怀疑论者会非常乐于同意，知识仅仅在于接受那些不可能出错的印象，但他们也接着认识到，根本不存在这样的印象。想想看，难道只是等到那个朋友走近些就足够了吗？万一要是长得完全一样的双胞胎呢？真是那样的话，即便走得再近一些，你也仍然无法分辨。好了，如果你很确定这位朋友没有双胞胎兄弟或姐妹，但这也还是不够，因为你有可能是记错了，或是你在做梦、喝醉了，甚或是出现了幻觉。如果明智的人要等到排除所有这些可能，直到自己的印象绝不可能出错之后才接受它，那么恐怕他是要一直等下去了：即便是最鲜明生动的印象也有可能出错。所以，斯多葛学派学者论证说，既然印象总是可错的，那么知识就是不可能的。

14

人们也许会注意到，这一立场的内在一致性是成问题的：既然学院派学者主张我们不能建立任何的确定性，他们又怎么能在“知识的不可能性”上保持如此确定的态度呢？类似的理论焦虑引向了更深层次的怀疑论。要理解这种深层的怀疑论，我们就要来设想这样一种独特的思维模式：它只包含**纯粹**的怀疑，完全不做任何正面的、建设性的断言，甚至也不能断定知识的不可能性。皮罗主义怀疑论就力图采用这样一种更为彻底的进路。这一学派以埃利斯的皮罗（约公元前360—前270）命名。对于皮罗这个哲学家，我们只是从其他哲学家和历史学家的叙述中间接地有所了解，而不是从他自己写作的文本中知道的，因为他几乎没有什么著作流传下来。青年时代的皮罗就参加了亚历山大大帝对印度的远征，据说他非常喜欢接触到的印度哲学。远征回来以后，皮罗就开始吸引一些追随者，最终变得非常有名。他的家乡还因此给他塑像，并公告说生活在这里的所有哲学家都可以免税。今天我们之所以还能受到皮罗的影响，主要是来自他的仰慕者恩披里柯（约公元160—210）的著作。正是他从一些古代思想资源中阐发出了怀疑论的观念，也就是今天被称作皮罗主义的怀疑论。

不论是学院派怀疑论还是皮罗主义怀疑论，两派共同关心的是“真理的标准”问题，或者说我们应该以何种准则来判别所要接受的东西，毕竟知识就要求我们不能随意地接受任何判断。不同的哲学家提出了许多不同的准则，以判别哪些才是值得接受的、最好的印象。既然知识要求我们做出深思熟虑的选择，我们也就不能随机地选择真理的标准。但要避免这种随意性，我们在选择真理的标准时就又需要某些准则。我们是否会用自己所偏爱的标准来为自身辩护？那看起来就像循环论证了。但如果我们寻找某个新的准则来支持我们的标准，那么我们就又需要更进一步的原则来为这个新的准则辩护，于是这就陷入了无穷后退。

皮罗主义者很喜欢把人们的注意力引向真理标准问题的上述疑难，但并不做任何肯定的主张说它永远得不到解决。在意识到上述困难之后，他们所采取的办法是普遍地悬搁判断。皮罗主义者发展出了一种普遍的策略，以便在所有主题上都产生怀疑：每当你试图在某一主题上做出决定时，你就想想这一决定的反面。你所应该做的并不是在这一主题上做出这样或那样的决定（那就会是“独断论”的立场），而只需要继续收集进一步的证据，在你的心目中使问题两方面的答案保持平衡。为了保持相反观点的平衡，有很多可以采用的技术。譬如，你可以设想一下，假如是在其他动物看来，或是从其他视角来看，又或在不同的文化中，事情会变得多么不一样。恩披里柯利用先前怀疑论者的工作，发展出一个扩充了的目录，其中详尽列举了各种能使你避免对给定问题做出具体回答的方法。他甚至还给出了一个用语的清单，全是怀疑论者会对自己说的话（“我不决定任何事情”；“或许是这样，或许不是”）。但恩披里柯并不是要把这些用语变成他自己教条的表达：他的怀疑论与其说是对实在的理论主张，倒不如说

15

是一种实践或生活方式。对所有问题保持开放或许听起来会更让人焦虑，但足以使人好奇的是，恩披里柯却报告说，他的印象是怀疑论实践给他带来了心灵上的宁静平和。（这当然只是一种印象——他不会确定地说自己达到了心灵上真实的宁静，也不会确定这是怀疑论所导致的，而不是随机的结果。）

怀疑论所面对的一种早已有之的批评，就是说它对人类的生存而言会是个问题：假如怀疑论者在“吃饭会不会解饿”这样的问题上都悬搁判断，那他们岂不是要面临饿死的风险？皮罗主义者回应说，行为的向导可以是本能、习惯或风俗，而不必是判断或知识：怀疑论者所要反对的是认知上的教条，他们并不需要反对自己的本能冲动或不自觉地产生的印象。怀疑论者当然可以放任地满足自己的饥渴之欲，但仍然不做任何关于实在的判断。

然而，“拒斥所有判断”并不是一种容易得到支持的怀疑论进路。在整个中世纪，西方哲学中具有主导作用的人物基本都是坚定的非怀疑论者。但在印度哲学传统中，怀疑论却得到了丰富和发展。其中，特别著名的是11世纪的印度哲学家室利曷沙的著作《驳论的妙诀》，它教给读者某些令人叹为观止的论证技术，以驳斥做出任何正面主张的理论，而不管这些理论的具体主张是什么。室利曷沙向他的读者承诺，只要运用这些怀疑的方法，特别是关注我们所用术语定义方面的困难，那么就能在论证中享受到“攻无不克的愉悦”。与古希腊人一样，室利曷沙也强调事物表象的欺骗性，因此我们的力量不足以发现事物的真实本性。他看起来也与恩披里柯一样，希望通过拒斥任何获取知识的积极努力，以便得到心灵的宁静平和。

对知识可能性的彻底怀疑在接下来的几个世纪中时常浮

现，特别是在人类智识方面经历剧变的时期。例如，16世纪正是新科学的成就挑战中世纪世界观的时代，那时的欧洲就出现了怀疑论的复兴。恩披里柯的著作被重新发现，他的论证得到一些哲学家的热情追捧，其中就有蒙田（1533—1592），他的个人箴言“我知道什么？”就表达了对皮罗式怀疑的热情。这种怀疑论的氛围不断蔓延开来：到17世纪早期，勒内·笛卡尔（1596—1650）写道，怀疑论远远不是一个已经消亡了的古代观点，而“在今天正生机勃勃”。《第一哲学的沉思》是笛卡尔最著名的著作，其中除了一些关于梦境和幻觉的令人耳熟能详的古代怀疑论论证之外，还对理性的局限性提出了一些全新的怀疑论论证。笛卡尔最深刻的怀疑论论述是，设想一个你被强大的恶魔欺骗的情境：那个恶魔在任何方面都致力于欺骗你，不仅给你输送虚幻的感官印象，而且即便是在你做简单的数学计算等抽象判断时，它也会让你的思维陷入歧路。如此鲜活生动的图景常常萦绕在哲学想象中挥之不去，尽管笛卡尔本人认为我们拥有摆脱它的确定方法。笛卡尔自己并不是怀疑论者，虽然他以其杰出的才能提出了不少怀疑论论证，他的真实目的却是要无懈可击地证明怀疑论是错的。但他对解决怀疑论问题的这份乐观并没有得到广泛的认同，近代早期的主要思想家仍然在坚持不懈地与怀疑论作斗争。第三章会更详细地考察笛卡尔及其论战对手的观点，在其知识实证理论的语境中理解他们如何处理怀疑论。

旧问题的新回应

20世纪以来，怀疑论的老问题出人意料地获得了某些全新的回应。英国哲学家G. E. 摩尔在1939年的一次公开演讲中提出了一个特别简单的方案。摩尔只是举起了他的手，然后说“这是一只手，而这是另一只”，以此来回答“如何证明外部世界的实在性”问题。他解释说，他的这两只手就是外部世界的对象，因此逻辑上就可以得出结论：外部对象是真实存在的。摩尔认为这是完全令人满意的证明：大前提是他有手，小前提是他的手是外部对象，或者用他的话说是“空间中能碰到的东西”，那么显然就能得出外部对象存在的结论。怀疑论者当然还会抱怨说，摩尔并不真的**知道**他有手——但也恰恰是在这里，摩尔提出应该把论证的负担甩给怀疑论者的一方。“如果你要说我并不真的知道，只是相信有手的存在，而实际上可能并没有手，这该是多么荒谬的事情！”摩尔坚持认为他知道自己有手，但并不试图去证明自己是对的。在揭露了怀疑论焦虑的荒谬性后，摩尔意在解释他不去证明自己有手的理由，也就是解释为什么我们应该认为他在这一点上拥有真正的知识。

首先，摩尔评论说，他主张无须证明而知道自己有手，并不等于主张任何人都**永远不能**证明他有手。事实上，他很乐意承认，在某些特殊情况下，人们或许能合理地证明自己有手。譬如，假如有人怀疑你是截肢患者，所佩戴的不过是人造的义肢，但实际上你并不是这样的，那么你就需要让他来仔细检查你的手，以便驱散他的怀疑。如果十分热切地想要证明这一点，你甚至可以让他感受一下你的脉搏，或是用某个尖锐的物体扎一下你的手。但是，不论这些旨在打消义肢疑虑的策略多么有效，摩尔还是认为，不可能有什么能够证明你的手存在的全能策略，就好像能够提供一个普遍的证明，以便能够打消所有可能的疑虑那样。因为可能产生疑虑的范围实在是太大了。举一个很简单的例子来说。假如有这样的打消所有可能疑虑的普遍证明，那么它就首先要证明，你不是一个睡着的截肢患者，在某次失去手臂的事故之后正躺在医院里做梦。尽管这种情况看上去令人难以置信，但我们是否能够证明这种情况并没有真的发生呢？对此问题摩尔依然持悲观的态度。然而，摩尔认为知道自己有手并不需要能够证明这一点，所以他也认为，即便你不能证明自己现在正在做梦，这也并不会阻止你知道自己并没有在做梦。在这里，摩尔再一次延续了对自己拥有知识的信心，尽管他也承认了在证明的能力上存在诸多制约："我在断定自己不是在做梦这方面无疑具有确凿的理由；我有确定无疑的自己是清醒的证据，但这并不等于说我有能力证明它。我没法告诉你我所有的证据，但为了给你一个证明，我至少应该要求做到这一点。"

既然摩尔主张有确凿的证据相信自己的清醒，他就拒斥了怀疑论推理所假设的力量。怀疑论者会说，**假如**你是在做梦的话，你就不可能通过看到两只手而知道你有手。对于这一点，摩尔事实上是赞同的，但他所要提醒我们的是，怀疑论的所有论证都落在那个大大的"假如"上：正如摩尔所言，对于那些知道自己并没有在做梦的人来说，无论他是否能证明自己没在做梦，他都不应该被怀疑论的焦虑所困扰。

然而，宣称自己可以有知识而无须证明的策略或许也在某些方面敲响了警钟：摩尔是否拒绝与怀疑论者斗争而径自宣布了胜利？摩尔愿意构建他所认为的"外部对象的存在"这一论断的绝好证明，却根本拒绝证明另一个类似的断言："那些手是存在的。"这难道不也看起来有些奇怪吗？更何况这两个断言原本就有着重要的差异：前者是哲学上的一般判断，后者却是日常意义上的个别判断。如果我们要支持某些普遍的哲学判断，显然就必须经过公开的推理与证明：譬如，我们对“外部对象”的意义可能就需要相当长的推理论证过程，而事实上摩尔在讲演中也的确在这方面做了大量细致的讨论。与此相比，像“这是一只手”这样的判断是如此基本，很难说还有什么更简单、更容易获知的判断来支持它。（数学上也有类似的例子，某些基本的判断被当作公理而不再需要进一步的证明。）如果怀疑论者试图颠覆这些基本判断的确定性，那么摩尔会建议我们，在放弃对原初常识的信任以前，我们首先不应该相信的反倒是怀疑论者臆想性的哲学推理。假如有人声称知道某个充满争议的哲学观点，却不能够证明它，那我们当然会合乎情理地怀疑他；但假如他声称的是知道当下环境中的某个简单的可观察事实，那我们也不应过多地反驳。

有的哲学家接受了摩尔的观点，认为怀疑论的论证中的确有某些错误之处；但即便是这些哲学家也不满意摩尔的解决方案，特别是他固执地坚持知识的常识观点。他们中的有些人就试图更精确地捕捉怀疑论的错误，同时也为知识的常识主张提供积极的辩护。摩尔的剑桥大学同事伯特兰·罗素就提供了一种主要的回应策略。罗素承认，怀疑论者说的有一点是对的：我们所有的印象（或用罗素的话说是“感觉材料”）都来源于不同于我们通常置身其中的实在世界的某种事物，这**在逻辑上是可能的**。但罗素对付怀疑论的策略是“最佳解释推理”，也就是即便我们承认这种逻辑上的可能性，也依然能够与怀疑论作斗争。因为他论证说，承认其逻辑上的可能性，并不意味着我们不能合

20

理地排除它，这二者之间有着巨大的间隙：在逻辑规则之外，我们还有更为狭义的合理性原则。具体说来，罗素所用到的就是简单性原则：在其余情况均同的前提下，简单的解释要比复杂的解释更合理。从逻辑上说，你通常归之于你的宠物猫的那些感觉材料——譬如喵喵叫的声音、猫毛的外观和质感等等——的确可能并不真的来自那只猫。或许它们来自一系列不同的动物，或是来自一系列令人费解的一致的梦境，又或是有其他诡异的来源。但在罗素看来，最简单的假设莫过于你通常会相信的那个：的确存在着这样一只猫，你跟它的定期互动让你产生了像猫的那些印象，萦绕在你的私人经验之中。对科学家而言，最合理的解释数据模式的方法，莫过于诉诸简单的定律；同样地，对我们来说，最合理的解释日常经验模式的方式，也莫过于承认一个有持存对象的简单世界，即"实在世界"的假设。

罗素的方案自然有它的吸引力，但也存在一些问题。因为，即便我们承认最佳解释推理是合理的策略，我们也依然会觉得它不够充分，因为它还是没有说明为什么我们由此得到的就是**知识**，而不只是合理的信念。把这一推理方式应用在其他语境中就会暴露出其潜在的弱点。例如，侦探在调查某项犯罪案件时也可以使用最佳解释推理。假设他发现管家的鞋子上有与案发现场相符的泥土，女仆的证词表明管家一直怨恨被害人，并且又在管家的床底下发现了一把有血迹的刀，那么侦探就能很合乎情理地得出结论：对这些证据的最佳解释就是管家实施了谋杀。然而，怀疑论者会指出，事情也许并不像表面上看到的那样。也许女仆才是凶手，而她只是非常巧妙地构陷了无辜的管家。这不是最简单的解释，但也许是真实的解释。假设侦探没有找到任何证明女仆卷入其中的证据，或许管家有罪就是最合理的结论，但并不能由此就说管家事实上的确犯了罪。同样地，怀疑论者也可以指出，尽管我们的经验很可能源自外部对象，或者说甚至我们也的确有理由相信是这样，但这仍然不等于说，这样的经验让我们获得了知识：在怀疑论者看来，知识需要满足比合理信念更高的标准。

罗素方案的更大麻烦在于，对我们的经验来说，实在世界的假设并不见得就比怀疑论者的解释更好。因为怀疑论者会论证说，假设欺骗性的恶魔存在，那么就能很巧妙地解释罗素所关心的那些经验特征。恶魔**当然**会给我们源源不断地输送生动而连贯的经验，因为恶魔一直努力欺骗我们，让我们相信存在着外部对象的世界。这种反驳也可以用在其他类型的怀疑论假设上。如果有人不喜欢恶魔假设中的超自然色彩，我们也可以替换成现代科学版本的怀疑论假设：只需假设你的大脑离开了身体而与一台超级计算机相连，这台计算机模拟某个连贯实在的经验，不断地给你输送感觉经验的信号；只要计算机的程序设计得足够好，能一直保持连贯性，并随着你的运动控制信号来调整显示的信息——譬如，如果你决定向左看，那么你的视觉经验也会随之而改变——那么，你作为一个钵中之脑而获得的经验，就与别人在物理环境中获得的经验，很可能仅从经验者的内在视角完全无法区分。所有你认为自己看到、感觉到的东西，譬如蓝蓝的天空、温暖的阳光，都可能是由那台超级计算机模拟出来的虚拟现实的一部分。假设这整个模拟的目标就是给你输送感觉经验，使之完全对应于你通常在物理世界中获得的感觉经验，那么，对最佳解释推理的支持者来说，还有什么理由认为实在世界的假设比钵中之脑的假设能更好地解释经验呢？这就是一个很有挑战性的问题了。而为了回应这一挑战，人们也提出了种种方案。譬如，美国哲学家乔纳森·沃格尔就论证说，实在世界的假设仍然是对经验更好的解释，因为它所需要的基本空间结构，较之于钵中之脑在实在世界中的虚拟现实的空间结构，要简单得多。

即便沃格尔是对的，这也不过是说，与其说我们是钵中之脑，倒不如说我们就生活在真实的世界中才更合理。但我们还是渴望一种更有力的方式来回击怀疑论。譬如说，有没有可能证明怀疑论论证中潜藏着巨大的困难？或者说证明怀疑论者在主张某些不仅不合理，而且更是完全错误的东西？近来，有的哲学家就试图采用语言哲学的工具，以这种更有攻击性的方式反驳怀疑论。这种新的反驳就是"语义学进路"，其背后的理论动机在于，通过考察语词获得意义及其与实在关联的方式，我们就能找到向怀疑论者开火的弹药。具体说来，这些以语言哲学构造的反怀疑论论证都出自语义外在论的思潮，这一思潮可以一直追溯到1960年代至1970年代的马库斯、克里普克和普特南。

语义外在论的主要观点是，语词之所以具有意义，并不是由于说话者将其关联到心理上的意象或描述（这被称作"语义内在论"观点），而是由于我们与周边世界的事物存在着因果联系。例如，在莎士比亚的时代，人们还把水看作一种基本元素；现代的科学家却用化合物H_2O来刻画它。即便"水是什么？"的问题在莎士比亚、现代科学家和街边的普通人那里得到的答案完全不同，他们却无疑都是在与同一种物质打交道。语义外在论者主张，之所以我们能够用"水"这个词指称同一种物质，正是因为这一语词的使用就建立在对这一具体物质的因果关联上。既然莎士比亚和现代科学家所看到和尝到的都是同一种液体，那么不管他们对这种液体的本质有何不同的认识，在使用“水”这个词时他们还是在指同一种东西。有时，相关的因果链条还需要经由其他说话者传递过来。譬如，今天没有谁见过那位法国皇帝拿破仑，但我们仍然可以谈论拿破仑，就是因为我们能够从其他来源那里获得“拿破仑”这个词的指称，当然还要再经由正确的因果链条才进一步回溯到那个历史上的拿破仑本人。因此，说话者可以对某件事情持有不同的甚或相反的观念，却依然能够谈论同一件事情，这在语义外在论的框架中就非常容易解释。例如，吉尔说拿破仑个子矮，比尔说拿破仑实际上还算当时中等偏上的个头，还知道所谓他小个子的谣言乃是出自其英国敌人之口，那么，他们关于拿破仑这个人的心理形象是很不一样的，但这并不妨碍他们所谈论的都是同一个人。

以语义外在论对怀疑论做出的著名回应，出自普特南1981年的著作《理性、真理与历史》。在这本书中，普特南批评的怀疑论者试图论证说我们可能只是钵中之脑。而普特南论证说，“我现在是钵中之脑”这句话，对于任何理解这句话意义的人来说都不会为真。按照他的理解，如果有这样一种生物，它所获得的所有信息都不过是来自虚拟现实装置的电信号刺激，那么它就不可能在与我们相同的意义上使用“钵”这个词。既然这种生物只与钵的仿真影像打交道，那么它所说的“钵”就不可能指称任何在这种仿真装置之外的物理世界的存在物。而我们对语词意义的把握却是在这种仿真装置之外，植根于我们和现实世界的“钵”打交道的历史经验。因此，我们之所以能够理解怀疑论者的假设，恰恰表明了我们正是生活于真实的世界之中：如果你实际上已经理解了“钵”是什么，那么你就不可能从来都是钵中之脑。

普特南的上述论证立刻引发了新的反弹。有的人就说，这只是深化了怀疑论的问题：现在我们应该担心的是，也许语词并不具有我们原来以为它们具有的那些意义。也有人注意到，即便普特南能证明，一个人如果总是钵中之脑的话就不可能谈论真实的“钵”，因为他从未有过与真实的“钵”打交道的经历，但普特南对于一个全新的钵中之脑依然束手无策。怀疑论者不只是主张你**现在和过去一直是**钵中之脑，有创造力的怀疑论者可以让你怀疑自己是不是从昨天半夜开始刚刚成为钵中之脑：你可以假设自己的生活直到昨天半夜之前都很正常，然后你在睡梦之中被一个疯狂的科学家窃取了大脑，使之与一台超级计算机相连接，进而模拟出一个虚拟的世界，还特别像直到昨天晚上之前你一直生活于其中的真实世界。假设这是一个非常完美的模拟，你当下的任何经验都无法排除它。那么，对普特南而言，非常不幸的是，语义外在论观点也必须承认你所说的“钵”这个词仍是有意义的，它的意义就来源于你先前在真实的世界中获得的对“钵”的过往经验。这个“晚近的钵”的图景仍可以意味着，所有当下的感觉经验都可能是幻觉（你以为外面是蓝天白云，实际上却是乌云密布，风雨交加，诸如此类），而这就是怀疑论者所喜闻乐见的恼人结论。

1999年的科幻电影《黑客帝国》把这个问题戏剧性地推向了更广大的受众。影片的主人公，由基努·里维斯扮演的“尼奥”发现，他作为一个1990年代办公室职员的生活只不过是虚拟的幻觉：真实的情况是，时间是两个世纪以后，人类已经在与

25

人工智能控制的机器的战争中败下阵来。这些机器将人类困锁在吊舱里，用大量同步的虚拟现实（“母体”）来给人以综合性的经验。尼奥发现了程序中的某些漏洞，在一些勇于反抗机器的人类的帮助下，他最终找到了逃脱虚拟现实控制的途径。此时他面临着两个选项：要么是在虚拟现实中继续过办公室职员的舒适生活，要么就是从母体中挣脱，参与反抗机器的地下运动。影片将这一选择的时刻戏剧化地处理为是吃红色药片（意味着母体之外的危险现实）还是吃蓝色药片（继续母体之中的舒适生活）的抉择。作为影片的主角，尼奥幸运地选中了红色药片。但令人迷惑不解的是，究竟是什么东西在此时利害攸关。经历一个真实的物理对象的世界，为什么就一定比沉浸于计算机模拟的图景中更好呢？

或许并不会更好。哲学家戴维·查尔莫斯坚决地为这样一种匪夷所思的观点辩护，他主张钵中之脑实际上要比我们所想象的好得多。具体说来，查尔莫斯试图证明，钵中之脑对其所处环境的日常信念也**为真**：例如，如果超级计算机让钵中之脑“看到”一本书，它因此会对自己说“我手里正拿着这本小书”，这里实际上并不存在任何欺骗或错误。巧妙的地方在于，对钵中之脑来说，它的语词和思想并不指称物理对象，而是指称由基本的计算属性所构成的对象。查尔莫斯也采纳了语义外在论的观点，将语词的意义与经验的原因联系起来。在这一前提下，钵中之脑所谈论的“这本小书”或“我的手”，实际上指的就是超级计算机中的子程序，以负责模拟对那些白纸和抓紧手指的相关感觉。这意味着，一个钵中之脑根据其经验的本质，可以**正确**地说它拿着那本书——因为，超级计算机中负责模拟对书的经验

26

的那部分程序，的确与负责模拟手的经验的那部分程序关联着，而且这种关联是以对钵中之脑来说是“拿着”的方式，这也是唯一一种钵中之脑所知道的“拿着”。从我们非钵中之脑的角度来说，我们当然会认为那个钵中之脑只不过是虚拟地拿着一本虚拟的书；但从钵中之脑自身的视角来看，它有理由认为自己正拿着一本书——它的语词具有与环境相适应的意义，且根据这些意义，钵中之脑所相信的东西是真的。实际上，没有什么东西能够阻止这个钵中之脑**知道**它正拿着一本（它认为的）书。因此，这个“怀疑论情景”并不会对我们日常生活的知识构成什么威胁。

按照这个思路，尽管虚拟现实从性质上看完全是计算的，但它并不会因此就不如那些由亚原子层次的基本粒子构成的物理实在。我们如果发现自己竟然是生活在这样的虚拟现实中，或许会让人惊讶，但如果查尔莫斯是对的，那么这种发现也不过是像发现量子力学或弦理论为真那样：一旦听闻物质的终极本性并不像我们所想象的那般，或许会暂时地让我们坐立不安，但它不应该动摇我们对普通知识的信心，譬如说我们关于鞋子、手或书的知识。（你当然会关心鞋子是否湿了或者穿着不舒服了，但你真的会关心鞋子最终是由什么东西组成的吗？譬如说究竟是由基本粒子，还是由某些振动着的一维对象，抑或是某个计算机程序指令构成的？查尔莫斯认为，答案很可能是否定的。）

怀疑论者原想用钵中之脑的图景恐吓我们，使我们相信并没有什么知识：“你甚至无法排除这样一种可能性：你的大部分信念都是假的。”与摩尔不同，查尔莫斯在这一点上同意怀疑论者：钵中之脑的图景的确无法排除，但他并不认为在此情形下大部分信念就都是假的。然而，我们还可能会为这一怀疑论图景

27 的某些变体深感头痛。像普特南一样，查尔莫斯也面临着这样的困难：如果主体是刚刚才沉浸在虚拟现实中的，那又该如何？在这一变体的怀疑论图景中，人们先前知道什么是从物理意义上“拿着”一本非虚拟的书。又由于他们所使用的语词意义就来源于其先前的经验，所以当他们刚刚被连接到超级计算机上，获得了对某个虚拟书籍的虚拟现实经验时，他们所说的话就会是假的：当这个新整合的钵中之脑说“我正拿着一本书”时，它所意指的是一本真实的、由基本粒子组成的书，所以它所说的话必定为假，也就不可能有知识。查尔莫斯非常清楚其中的困难，但他试图弱化其影响，办法就是指出那个新整合的钵中之脑仍然残存着它先前获得的知识。如果查尔莫斯是正确的，怀疑论者根本不可能轻易地构造一个全方位否定知识可能性的图景，即不可能证明**几乎所有**关于过去、现在和未来的信念都是假的。当然，顽强的怀疑论者还可以指出，否定知识可能性的目标只需通过一个温和的图景就可以实现，在其中你只是（例如）对当下感知的所有对象的判断都是错的。如果你刚刚被连接到一台超级计算机上，而它又不断地模拟着你的经验，那么即便是查尔莫斯也得承认说，你不可能知道自己是否正穿着鞋子。同时，你还可能继续持有下述疑虑：如果不能排除这一图景，那么你就不能主张说自己已经知道了某个事实，无论你是否身处虚拟现实之中。

那么，我们是否不可能从一开始就拒斥怀疑论？这或许是一个试图攻击怀疑论推理的非常有野心的计划。与此相反的是，有的哲学家主张某种更具防御性的进路。按照他们的观点，我们不要试图用怀疑论者的词汇来回答他们提出的问题，而是要努力从一开始就超出怀疑论者所能掌控的范围。既然怀疑论者的目的是质疑所有你所说的话，那么也就不可能在你和他之间找到一个共同的基础，并且还能在这个共同的基础上说服他相信自己错了。如果从怀疑论者所接受的前提入手，你就只会作茧自缚，还很难从他挖的坑里跳出来；而如果选择从常识的世界观出发，我们就很容易构建某些防御性的机制，以抵御怀疑论的魔咒，同时接受常识上的假设，即我们的确拥有很多知识。 28

一种拒斥怀疑论的方法是，对怀疑论观点的吸引力提出一种诊断：既然这样的观点会引向非常陌生乃至荒谬的结论，那么为什么它还让人那么趋之若鹜呢？在这方面，哲学家们提出了五花八门的观点。来自牛津的哲学家蒂莫西·威廉姆森提出，怀疑论尽管有着令人失望的结论，但它之所以一开始看起来很吸引人，是因为它本来是个好东西，只不过走得有点太过了。其中好的部分在于，它说明我们具备一种有益的批判性能力，以对自己所相信的东西做“再确认”（double-check），这就要暂时悬搁对它们的判断，以便看清楚它们是否与我们所知的其他东西相吻合。但这种悬搁个别信念的能力只是作为一种有用的免疫系统，以便清除那些不一致的或无根据的观念；而怀疑论则不然，毋宁说它是自身免疫性疾病，因为它的保护性机制太过了，以至于机体的健康部分也遭到了袭击。一旦我们悬搁了太多的东西——譬如说，一旦我们连整个外部世界的实在性都要怀疑——那么我们也就无从支持或再确认我们所相信的完全合理的东西。

还有一些其他因素也可能导向怀疑论。或许是由于知识的基本模型中存在某些错误，这就构成了怀疑论论证的基本背景。

按照这个得自斯多葛学派的模型，我们是从印象（或观念、感觉材料等）出发，然后才有一个接受或拒绝这些印象的环节。如果现实生活中被广为接受的印象与梦境或模拟仿真的效果并无差异，在我们看来完全一样，那么似乎就很难说清楚，我们究竟是怎样辨别出哪些是应该接受的东西的。在早期近代哲学中，这 29 个问题表现得尤为突出。

第三章

理性主义与经验主义

近代早期

在文学、音乐和建筑等许多领域里,“现代”(Modern)这个标签只能延伸到20世纪的早期。而哲学却与众不同,它的近代(Modern)时期早在400多年前就开始了。[①]哲学的这种特殊性是由于16世纪人们对自然的理解发生了翻天覆地的变化,这场巨变同时也使得我们对知识本身的理解发生转型。在这条线索上,近代意义上的思想家可以追溯到伽利略·伽利雷(1564—1642),他投身于与我们今天非常相似的研究计划。如果我们回望近代之前的时期,我们会发现很多异乎寻常的事情:那时人们对于自然如何运行,以及自然如何能被认识,还存有很多与今天非常不同的观点和看法。

① 原文中用的词都是Modern,一般以中文的“现代”来对译。但在西方哲学史的翻译中,通常译作“近代”。换言之,作者所试图做出的区分,在中文语境中已经以不同的译名区别开来了。——译者注

举一个近代之前奇特思维方式的例子，尝试理解一下文艺复兴时期的思想家帕拉塞尔苏斯（1493—1541）的下述论断：

> 整个世界围绕着人，就像圆围绕着圆心。由此可知，所有事情都与这一点有关，正像一颗苹果种子，被它外围的果实所紧紧围绕和保护着……天文学理论通过研究行星与恒星而理解的所有事情……也都可以应用到身体的宇宙之中。

30

这位传统的思想家把宇宙看作围绕人的东西，试图通过发现宇宙与我们之间的类比关系，来获得有关自然的知识。这也就是把实在看作我们在心灵中创作的带有象征意味的艺术作品。（见图3）

到了16世纪，这种把人视作所有存在物中心的观点受到了严重的挑战，主要是来自一些尚未得到解释的发现，尤其是尼古拉·哥白尼（1473—1543）的主张，即地球实际上并不是宇宙的中心。旧传统试图阻止新观念的产生。例如，伽利略发明的望远镜发现了木星有几颗卫星，而守旧的传统学者弗朗西斯科·希兹却论证说，那些观察结果显然是错的。按照希兹的观点，既然动物的头上都只有七个孔（两只眼睛，两只耳朵，两个鼻孔和一张嘴），金属只有七种，一周也只有七天，那么同样地也就不可能存在多于七个“漫游的行星”，或是任何除了恒星以外的天体。

希兹并没有赢得论辩的胜利。这当然并不只是说我们相信伽利略是对的，即太阳系中有超过七颗星体在运转。更为根本的是，我们对自然和知识采取一种完全不同的思维方式。我

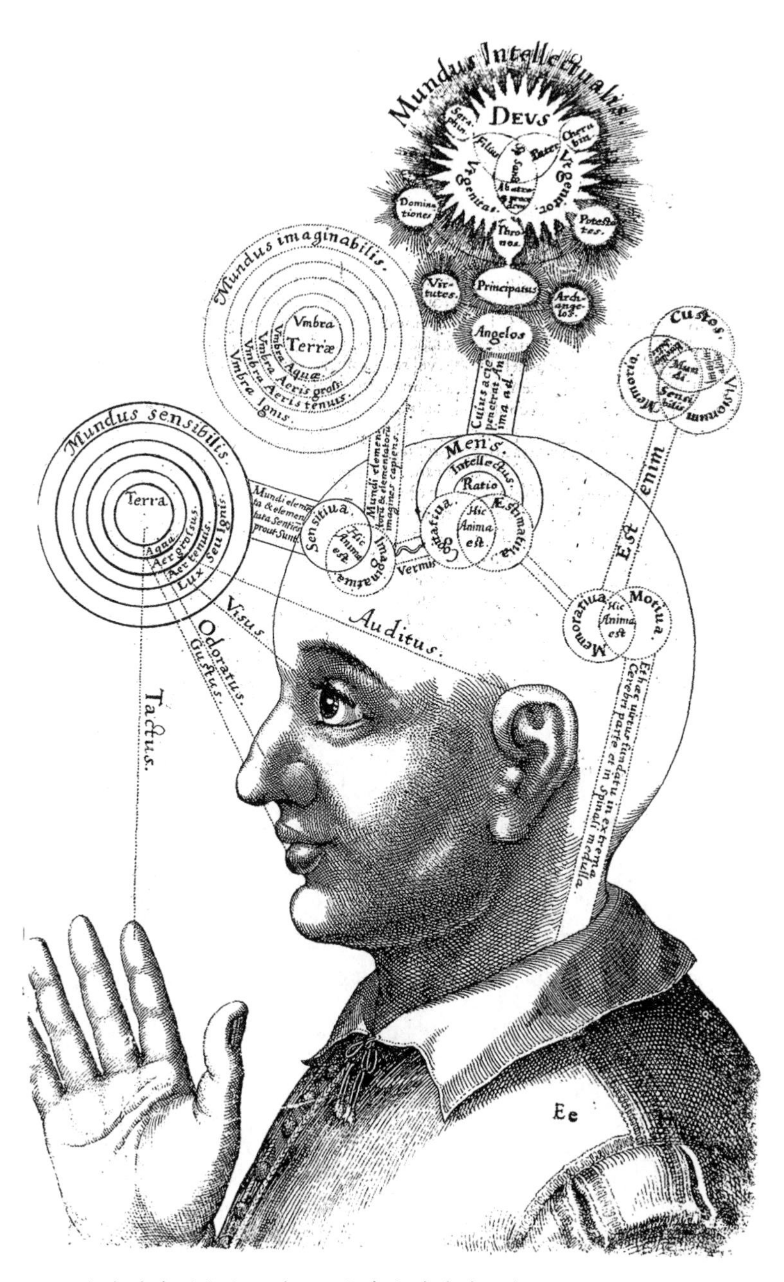

图3　帕拉塞尔苏斯主义关于人与宇宙关系的观点

们不再期待自然事实有任何特殊的人类意义——譬如，“为什么行星必须只是七个而不能是八个或十五个？”这样的问题。同时，我们也认为知识应当从对自然做系统的、持开放态度的观察中获得，而不是像希兹那样，求助于任何类比和模式。然而，向近代的转型并非易事。近代之前盛行的这种模式导向的思维方式，自然对像我们这样寻求意义的生物充满着吸引力。这样的思维方式也可以在不同的文化中找到。例如，在中国传统思想中，金木水火土的“五行”就对应着五种感觉，这也是在人类的内在领域与外部世界之间构建的类似对应。并且，更有吸引力的一点是，这种近代之前的观念还很容易适应我们日常的感觉经验：不加反思的话，地球的确看起来是固定不动的，而太阳却每天在空中东升西落。而你要说服自己相信一个数学上更简单的模型——譬如以太阳为中心的太阳系模型——才是正确的，反倒要花很多工夫去证明。

在短暂的文艺复兴时期，新旧两种思维方式彼此博弈，争夺主导权。正是两者的对抗让不少哲学家倒向了怀疑论的立场。旧理论把地球置于宇宙的中心，新观念则要使地球围绕太阳运转。正是看到了这两种观点的激烈碰撞与冲突，蒙田才下决心主张说，唯一明智的道路是不支持其中任何一种观点。但也并非所有哲学家都乐意置身事外，其中的一些人就选择为新的科学思维方式而战。

勒内·笛卡尔的理性主义

笛卡尔（图4）1596年出生在法国，接受的是传统的耶稣会教育，专心研究亚里士多德及其中世纪阐释者们的古典思想。

图4　勒内·笛卡尔，1596—1650

当他后来接触到研究自然的新兴道路时，他才对自己曾经在学校里学到的东西有了新的反思。他的《第一哲学的沉思》（以下简称《沉思》）开头即坦承，自己从孩童时代起就摄入了“非常多的虚假观念”。他接着写道：“我意识到如果我还想在学问上有任何建树，并且能够稳定而持续下去的话，我就有必要在我的生命里彻底地摧毁所有自己相信的东西，从基本的东西重新开始。”笛卡尔在《沉思》中提出的摧毁性纲领就是第二章提到的系统性怀疑论的方法，从对错觉经验的微小怀疑出发，最终上升为一个彻底怀疑所有外部世界的恶魔图景。这一图景出现于笛卡尔书中写到的第一个沉思末尾，作者宣称他已经成功地把所有固定的观念擦除干净了：他已经怀疑了所有先前自己持有的观念。33

在第一个沉思之后的五个沉思，都致力于仔细地重建一种新的世界观，笛卡尔希望为此提供不可撼动的牢固根基。第一个能够确定无疑知道的真理乃是我们自己的存在：无论是否有一个恶魔一直在欺骗你，至少有一个事实你是不可能被欺骗的，那就是你自己的存在，你是一个思考着的存在者，在你思考自己的时候你就是存在的。34 到这一步是没什么问题。但如果“我存在”只是用来支持其他知识的非常单薄的基础，笛卡尔采用了一个有意思的窍门，试图从中得到额外的收益。他在第三个沉思中提出了一个反思性的追问：究竟是什么使他知道了这第一个事实？他说，那只能是“对我所断言之事的清晰而明确的知觉”。如果这是对的，那么我们就应该可以信任这样一条法则：“凡是清晰而明确的东西即为真”——笛卡尔推论说，因为如果这条法则是可错的，我们甚至就没有办法主张知道我们自己的存在。第三个沉思于是就大胆地为这条普遍法则辩护。

由于笛卡尔只确立了他自己的存在，所以他能用上的资源实在有限。但他继续考察自身，于是发现在他的心灵中有着各种各样的观念，因而他就开始清点这个观念仓库的存货。笛卡尔没有去考察自己的观念是否都反映现实，而是注意到他的这些观念之间存在着显著的差异。它们所表征的对象类型不同，有天使、人类、太阳、天空和上帝等等；从表面上看，它们也有着不同的来源。有的来源于外部对象，譬如相信壁炉能取暖的观念；有的则似乎是被发明出来的，譬如他对一种希腊神兽的观念，那是一种有着鹰的脑袋的飞马；而还有一些则似乎是被植入到他头脑里的观念，或者说是与生俱来的，譬如关于真理的观念。只要笛卡尔还有恶魔图景的顾虑，怀疑有个恶魔在任何情况下都会欺骗他，那么他就不可能确信自己的观念真的具备它表面上的来源。但笛卡尔最终发现了往前更进一步的道路，认识到那些表面上看似不同的观念实际表征的是更大更少的东西。在所有观念之中，有一个观念居于其他观念之首，即他对上帝的观念，对一种无限完满存在的观念。笛卡尔推论说，完满的观念不可能从任何不完满的来源中获得，所以它必然只能来自上帝，而且必须是天赋内在的，是我们与生俱来的观念。笛卡尔之所以能从上帝的观念来论证上帝的存在，其实是借助了一个因果原则：即便是纯粹观念上的完满也不可能由任何不完满的来源所导致，而并非所有读者都会完全信服这一原则。但是，一旦确立了上帝的存在，笛卡尔就很快转而论证说，我们所有清晰而明确的观念——譬如对真理、数字和纯粹几何形式的观念——都是天赋观念，完全值得信赖。因为，完满的因此也是仁慈的上帝不会把任何欺骗性的观念赋予我们的头脑。

如果我们那些清晰而明确的天赋观念都是绝对可靠的，那么我们何以还会犯错误呢？显然我们经常出错，所以要解释这一点就成为有意思的问题。笛卡尔的下一个任务，就是要把错误的来源与我们那些值得信赖的清晰明确观念区分开。笛卡尔说，我们的有些观念乃是模糊而非清晰、混乱而非明确的。那些朴素的感觉经验具有一定的意义，例如饥饿、口渴，或是某些对颜色与气味的感觉，但这些观念并没有向我们展示事物自身具有的真实本质。譬如，在昏暗的光线中，你可能搞错了某个东西的颜色，因为颜色的观念并不清晰而明确。相反地，你对数字5或三角形的观念却是清晰且明确的，因为它们不会不向你展示其所表征对象的内在本质：你完全不用担心对数字5的观念错误表征了那个数。这里笛卡尔实际上回应了斯多葛学派的思路，主张如果我们基于那些模糊且混乱的观念做出判断，那么我们就不得不面临出错的风险；而只要我们坚持立足于那些清晰且明确的观念，我们就不可能出错。所以，如果我们想要获得关于像颜色和气味这样混乱的事物的知识，那么我们就应该以明确的数学或几何术语去分析它。这其实正是我们现代人做的事情，我们用分子的形状来解释气味，用可测量的光的波长来解释颜色。

这就是所谓的理性主义观点，它把抽象的概念置于寻求知识的核心。笛卡尔对代数和几何都做出了很大的贡献，特别值得一提的是在几何学上还发展了笛卡尔坐标系。所以他也非常坚持用数学工具分析自然这一现代性的事业。然而，在解释人类错误的来源方面，笛卡尔仍然面临着诸多困难。这种困难恰好来自他对下述主张的依赖：我们的天赋理智结构来自某个完满的上帝。在第四个沉思中，笛卡尔主张上帝不应该为我们的错误负责，因为这些错误乃是我们自由的后果——具体说来，就是人们自由地接受了某些不那么清晰与明确的观念。自由当然不是什么坏事，如果它被滥用，那也是我们的过错，不关上帝什么事。然而，在这一点上仍然存在一个谜团：为什么上帝在给予我们极好且完全精确的数学抽象观念之外，还要赋予我们一些

36

模糊与混乱的身体感觉呢？直到他的第六个也是最后一个沉思中，笛卡尔才解决了这个与我们的感觉相关的困难课题。

笛卡尔指出，我们的心灵有着诸多能力，既有理智又有想象力。我们的理智负责代数或演绎几何学等抽象操作，而想象力则与此完全不同。为了领会两者之间的差异，笛卡尔要求读者思考一个千边形—— 一个有着1 000条边的几何图形。从理智上说，你完全可以做到这一点，能看出千边形和1 001条边的几何图形间的不同，你也能在正确公式的引导下做计算，得出结论说千边形的内角和等于1 996个直角的度数。但是，从想象力上说，情况就困难多了：笛卡尔这样预测，假如你想要想象出千边形究竟是个什么模样，你肯定会得出某些非常混乱的结果，譬如得出了某个有许多条边的图形，又很难与其他也有很多条边的图形区分开。于是，笛卡尔的结论是，想象力并不只是从理智中读取那些清晰而明确的观念，而是导向了其他的东西——身体性的感觉。

笛卡尔最终的结论是，我们的感觉及与之相关的想象力，目的并不在于向我们展现事物自身的本质，那其实是理智的任务；相反地，感觉服务于身体与灵魂。我们对饥饿、疼痛、气味和颜色的感觉帮助我们的机体正常生存，这本身当然也是仁慈的上帝希望保全的东西。如果我们时刻想着，感觉并不是用来展现事物真实本质的，那么我们就能够避免感觉的误导。确切地说，感觉所提供的东西应该根据我们清晰而明确的观念做出检验与解释。通过观察经验中的一致性与融贯性，我们就能够区分真实的经验与非真实的梦境，尤其又因为一个仁慈的上帝不会愿意让我们终生深陷于梦境之中。如果我们能够小心地协调运用心灵的各种能力，让我们混乱的感觉从属于天赋理性的规训，那么知识就会是可能的。

笛卡尔对于自己的理性主义体系非常乐观，在其前言中主张自己的论证“乃是这样的类型：我认为任何人类的心灵都不可能再发现有比它更好的了”。然而，他的这种热情并未得到广泛的认同。很少有批评者怀疑他对自我存在的确定性主张，然而除此之外，他的论证遭到了非常多的反驳。神学家安东尼·阿诺德立即指出，笛卡尔在第三个沉思中的论证存在某个循环：当他要论证“清晰而明确的观念是值得信赖的”时，他的论据仍然来自那些清晰而明确的观念。这个问题后来也被称作“笛卡尔主义循环”，直到今天仍然有损于笛卡尔在哲学家中的声誉。波希米亚公主伊丽莎白指出，对于理智的灵魂如何能够与物质的身体交互作用的问题，笛卡尔远远没有给出令人满意的答案。英国哲学家托马斯·霍布斯追问笛卡尔：“天赋观念”究竟是怎样起作用的？最后这一批评被另一位英国哲学家约翰·洛克（1632—1704）往前大大推进了一步。

约翰·洛克的经验主义

当洛克（图5）还是牛津大学一名人文学科的本科生时，他也研读了一些传统哲学的内容。但只有到毕业多年以后，他读到笛卡尔的新哲学时，才真正对这一主题产生了兴趣。洛克兴趣广泛，除了知识论以外，他还研习医学，更是一名成功的政府官员。但就是在这诸多兴趣和追求之中，洛克坚持发展了一种与笛卡尔相对立的知识理论，并花了二十年时间写作他的主要著作：《人类理解研究》（*Essay Concerning Human Understanding*，以下简

图5　约翰·洛克,1632—1704

称《研究》)。洛克在《研究》的导言中写道,此书的核心目标就是"厘清意见与知识的边界"。洛克相信,以他所谓"历史的平常方法"研究人类心灵的自然运作,就能够找出人类所能知范围的边界。他主张,在能知的边界之外,人类完全可以具有信仰或意见,但他们不会再有确定性,也不应该批判攻击那些持不同意见或信仰的人。(在洛克的计划中,支持宗教宽容乃是厘清知识边界的重要内容。)

洛克用"历史的平常方法"来挑战理性主义者对天赋观念的诉求。回顾一下人们实际上是如何做出判断的,理性观念真的从一开始就在我们的头脑之中了吗?理性主义者主张说,某些原则基于全体人类,因而必然是天赋的。譬如,"存在者存在","同一个东西不可能既是这样又不是这样",等等。但洛克基于自己对人类的观察提出了不同看法。"孩童当然知道某个陌生人不是自己的妈妈,知道奶瓶不是一根棍子,要远远早于知道'同一个东西不可能既是这样又不是这样',这难道还会有什么疑问吗?"所以,洛克非常反对说这些原则是孩童也具备的天赋观念,只不过还没有被注意到。他宣称,"如果说某些真理已经赋予灵魂之中,却未被感知或理解,那简直近乎于矛盾了"。如果说,"赋予灵魂之中"并不意味着灵魂觉察到了某些东西,那还能意味着什么呢?天赋观念论者试图主张,孩童只有能成熟地运用理性之后,才会认识到那些早已赋予在灵魂之中的真理。而洛克的反驳是,既然真理的发现需要运用理性能力,那么那种天赋的真理看起来纯属多余。这就好比是说,太阳也是上天赋予我们之中了,只不过它还需要视觉能力的发展,才可能被心灵注意到。

理性主义者主张心灵被预先设定了某些天赋观念，而洛克却主张我们最早具有的观念都来自感觉：在感觉开始标记之前，心灵就是一张“白纸”。在经历重复经验之后，孩童渐渐能够识别出物体与人，就是原初的“感觉观念”集合成了不同的模式。随着心智能力的发展，我们也就从对心灵自己运作的观察中发展出了洛克称为“反思观念”的东西。这样两个来源，一个是对外部对象的感觉，另一个是对心理活动的反思，就构成了人类获得简单观念的终极源头。在洛克看来，简单观念从这两个源头中形成，进而成为构成所有人类思想的最基本单元。这种以经验为中心解释知识的理论进路就是经验主义。

经验主义知识论虽然主张思想的原初材料来自经验，但它也为经过处理的其他材料预留了空间。譬如，我们先是接受了来自经验与反思的简单印象，然后以某种新的方式把它们联结起来，形成某些更为复杂、精致与抽象的观念，例如正义、财产和政府，或是幻想出某些虚构的生物以及新的发明，等等。一旦我们从更简单的成分中组合出了那些精致观念，我们也就能够具备关于它们的真理性知识。例如，洛克认为“无财产的地方也就没有不公正的情况存在”，这就是一个“具有欧几里得几何证明的那种确定性的命题”。如果你不能直接看出其中的缘由，这可能是由于你对“财产”和“不公正”的观念不同于洛克的观念。洛克把财产定义为“对任意事物的权利”，而把“不公正”定义为“对那种权利的侵犯与破坏”。所以，如果你按照他对这些术语的关键定义，你就很容易得出“无财产便没有不公”这样的结论。

当然，洛克也意识到，我们实际上并没有以相同的方式定

41

义语词。正是由于人们常把不同的观念关联到所使用的语词之上，所以在道德与政治议题上才会产生那么多恼人的争论，而实际上这些争论并非真的出自观念上的深层差异。如果我们能够对自己所使用的语词意义有非常清晰的理解，譬如说能够把它们分解为更简单的词项，那么我们也就能够解决这些单纯的语词之争，从而能够达到更大的共识且获得更多的知识。

然而，即便我们非常小心而精确地定义了语词，我们也还是发现，知识是有边界的。洛克对知识的定义是："对我们任何观念间的关联与一致，或分歧与矛盾的知觉。"观念间的一致或分歧可以有很多种方式：洛克称之为知识的"程度"。

最明确程度上的知识是**直觉知识**（intuitive knowledge），这也是我们对观念间的一致或分歧的直接把握。例如，"圆不是三角形"。稍微难一点的是**演证性知识**（demonstrative knowledge），心灵从中也看到了观念间的一致与分歧，但必须要借助于某种连接观念的链条：譬如，三角形的内角和等于两个直角，这个知识就是演证性的，因为它需要经过一系列证明步骤才能得出。

42

最弱程度上的知识被洛克称为**感性知识**（sensitive knowledge）。感性知识区别于其他知识的地方在于，它所呈现的并非观念间的普遍真理或关系，而是我们所体验到的具体对象的存在。按照洛克的观点，你对感觉对象的存在就具有感性知识，譬如，你现在正穿着的衣服。洛克称，如果你的确感受到了某个东西，你就会得到与仅仅是回忆到它或想象到它都不同的经验。譬如，你在白天真实地看到了太阳，而到了夜晚只能回想起太阳的样子，你完全不可能忽略两者之间的巨大差别。由于感性知识是把我们的观念关联到实在上，因而它似乎脱离了洛克的知识定义，因为按照定义知识只能是对观念间的一致与分歧的知觉。但是，只要我们还记得，观念也包括反思观念在内，由我们对心智运作本身的觉察产生，那么我们也就可以把感性知识看作对观念间一致性的感知。因为，在任何感性知识中，譬如你知道自己正穿着的衣服是存在的，你体验到你自己对衣服的感觉的心智运作，会区别于当你仅仅回想起或想象着自己穿着衣服时的心智运作。在真实的感觉中，你能识别出它与感觉观念的一致性，这是一种你在触碰面前真实存在的东西的观念，它远远不同于你对回忆或想象等心智运作的观念。

洛克并没有在回应怀疑论方面花费太多精力，而恰恰是怀疑论者会担心说当我们觉得自己有真实经验的时候其实并没有体验到实在。洛克不在意怀疑论，可能是由于他认为任何人都不可能做一个真诚的怀疑论者。洛克认为我们已经“不可改变地意识到了”我们所感知事物的实在性，最多只能假装对它持怀疑态度。同时，洛克非常强调知识对实践目标的意义：如果那些看似真实的东西已经真实到足以成为快乐与痛苦的可靠来源，那么你就能确定无疑地相信它的存在，而这完全就已经是你足以应付实践目标的信念程度。对洛克而言，知识归根结底是追求幸福的工具或手段。

但知识并非唯一的向导。洛克论证说，我们有很多行动并非由知识主导，而是由某些更弱的判断来主导。在他看来，判断并不会给我们任何确定性，但它仍然可以使我们主张某个论断很可能为真——并且，在众多情形中，这里的概率会足够高，能够使我们获得实践上的确定性。例如，我们从他人的证言中获得的信念，在洛克看来就只是某种判断，而并非真正的知识（这其实是一个很有争议的主张，我们会在第六章再次讨论它）。

洛克的经验主义与笛卡尔的理性主义都是在近代早期涌现出来的对自然新的思考方式，尽管它们分别强调了这一思路的不同侧面。笛卡尔侧重于数学与抽象观念的重要性，洛克则重点关注经验与观察的意义。既然这两位思想家都主张近代早期的新科学思想，为什么他们还会提出如此不同的知识理论？一种可能的理由是，他们是从不同的角度来批评知识问题的。在《沉思》中，笛卡尔采取了一种第一人称视角。他的问题是："我能确切地知道什么？"从对自身存在的内在意识出发，笛卡尔通过其心灵的能力与内容逐渐向外扩展。从第一人称视角出发，最清晰的知识乃是那些完全不需要依赖于外部环境的知识：即便我们对外部世界缺乏确定性，我们依然能够对抽象观念的纯粹理性运作保持充分的信心。洛克也带有笛卡尔的这种内省性倾向，但他在很多论述中也同样愿意采取一种第三人称的视角，借用他对他人的观察结果。对洛克而言，要回答的最主要问题是："人们究竟知道什么？"孩童如何认识到奶瓶与棍子的差异？从第三人称视角来说，我们自然地会承认，对奶瓶做出反应时，那个婴儿的确体验到了某个实在。因此，在第三人称观点中，感性知觉就是个人与其周边环境的关系，这构成了知识的自然基础。

当然，理想的知识理论是要能同时解释抽象的东西与观察性的东西，以及两者是如何配合的。而取得这一方向上的进步有赖于对第一人称与第三人称观点间联系的深层理解。这个主题在当代知识论中依然是非常活跃的研究方向，我们将会在

第五章中详加讨论。按现在的术语来说，在第一人称与第三人称观点之间的选择，就是在“内在论”与“外在论”之间的选择。自1960年代以来，这一选择的重要性变得非常清晰，那时哲学家们正费力回应对知识概念的传统分析提出的挑战。 45

第四章

知识的分析

盖梯尔的挑战

在一个周日的下午，史密斯走过一个废弃了的火车站，他意识到自己忘了时间，于是瞥了一眼火车站的钟，发现它清晰地指向一点十七分，而这正是当时的时间（见图6）。你可能会想说，史密斯现在知道了时间是一点十七分，但先不要着急，这里还有个额外的事实：那个钟实际上已经坏了，它的表针已经两天没动了。之所以史密斯看到这个钟时它没给出错误的时间，完全是个巧合。那么，通过看一个坏了的钟，史密斯真的能**知道**现在是一点十七分吗？许多人会想说，“不，他并不真的知道”。如果你并不持相同看法，尝试看一下方框4中的事例是否能更好地帮助你。

伯特兰·罗素在他一本出版于1948年的书中提出了那个“停走的时钟”的例子，以此作为真信念并非知识的例子。罗素还提供了一些其他的例子，譬如某个买了彩票的人热切地相信

图6 停走的时钟

自己会中奖，而且实际上也的确中了奖。这在当时并未引起多大的注意，但在“停走的时钟”例子里，的确还隐含着某种特殊的东西。十五年之后，当哲学家们再次审视这些例子，却发现了某个非常有趣的特征。在“停走的时钟”例子里，史密斯与那个 46

非理性地乐观相信自己会中彩票的人不同，是一个完全以合乎情理的方式做判断的人：他对自己有关时间的真信念拥有某些看起来非常值得尊重的完美证据。通常说来，如果有人问你是否知道现在几点了，你可能只是扫一眼自己戴的手表，然后就会[47]直接告诉他。你不会说："我不能确定自己是否**知道**时间，因为我还需要再次确认我的手表运转正常。"

促使哲学家们重新思考"停走的时钟"例子的，乃是一篇发表于1963年的短文，作者是美国哲学家埃德蒙德·盖梯尔。盖梯尔的目标是挑战分析知识的传统方式。在那个时代，知识论的主要观点是，知识就等同于得到辩护的真信念（这个等式也被称为"知识的经典分析"）。当然，那时在一些关键概念的理解上还存在争论。譬如，辩护是否要求某种特殊类型的证据？或者是否要求具备面对挑战来捍卫某一论断的能力？但除了这些细节上的争议以外，经典分析的基础结构从未受到真正的挑战。然而，盖梯尔在他1963年的文章中，用两个很像"停走的时钟"的例子来表明，即便某个判断为真且得到了辩护（或者说得到了证据的合理支持），它也仍然有可能不是知识。哲学家们很早就认识到，并非所有真信念都是知识，甚至柏拉图就讨论过凭运气猜想得到真信念的问题。而盖梯尔的观点提供的新内容在于，即便是得到辩护的真信念也依然可以不是知识。

如果那个人凭借停走了的时钟具备了真信念并获得辩护，却没有知识，那么经典分析必定错失了某种东西。假如信念的辩护与其为真之间并无恰当的联系，那么仅仅把辩护加诸真信[48]念之上，并不足以产生知识：知识乃是某种大于真信念与其辩护之和的东西。

方框2　知识的经典分析

（令S为某个人或认识主体，*p*是一个命题）

S知道*p*，当且仅当：

（1）*p*为真；

（2）S相信*p*；

（3）S的信念*p*得到了辩护。

初看起来，经典分析的问题似乎很容易找出来。在“停走的时钟”例子中，你或许会注意到这样一个特征：史密斯似乎持有某个假信念，他想当然地认为钟是在运转的，而这是个错误。那么，如果我们在经典分析上加一条额外的限制，不允许借助假信念来通达真理，这样是否就能够保留住经典分析的基本观点了呢？或许，获取知识正像过一座木桥，如果碰巧踩到一块烂透了的木板，你就全然无法达到目的。有这样一条避免依赖假信念的规则，就能保证你获取知识的每一步都走得稳当。

于是，在盖梯尔文章发表后的几个月之内，这个排除假信念的论证策略就被提了出来，然而又很快被否决了，基于两方面完全不同的理由。一方面，即便的确借助了假信念的支持，至少表面上看人们也**能**具有知识。设想一下，有个侦探正调查一起光天化日之下的袭击案，他走访了十多个目击者，都一致说看见了琼斯袭击史密斯；他又把众多物证收集起来，包括从琼斯的指关节上采集的符合史密斯血型的血迹，得到史密斯的陈述，甚至还有琼斯自己的供词。那么侦探现在知道是琼斯打了史密斯吗？当然他是知道的。（只有怀疑论者会有不同答案。）但现在假设，在侦探走访的那十几个目击者中，有一个谎称自己看到了袭击（其他人都是的确目击了袭击且诚实地提供证词）。如果侦探相信，所有证人**包括那个说谎者在内**都看到了袭击的发生，那么对于“琼斯打了史密斯”这个判断，他就是在用一个假信念提供支持。然而，这似乎并不能说，此假信念就足以阻止侦探对此拥有知识，因为除此之外他所拥有的其他确凿证据也早已足够了。侦探的这唯一的假信念就像是桥上唯一一块烂透了的木板，但桥被其他的好木料加固了，于是也能够安全通过。因此，我们不会也不应该赞同这样简单地否定知识，就好像只要有假信念的存在就完全不会再有知识一样。

49

还有另一些试图拯救经典分析的窍门。例如，我们可以主张那个侦探在本质上并没有依赖说谎者的证言：即便没有那个说谎者，他也依然能得出相同观点。或许，我们可以说，有知识的人不能在本质上**依赖**任何假信念，由此捍卫经典分析的立场。打个比方说，就像是过木桥时，烂透了的那块木板不能是实质性承重的那一块。但是，在形成判断的过程中，如何才叫作“在本质上依赖于”某物，却又是非常难说清楚的事情。很多理论都曾尝试说明这个问题，但都无济于事。同时，想要通过排斥假信念来捍卫经典分析的整个论证策略，也逐渐丧失了哲学家们的热情支持，因为这一进路在其他的前沿讨论中也落败了。尽管盖梯尔自己举的例子的确包含了对假信念的依赖，但是，人们后来发现，即便不包含任何假信念，那种有得到辩护的真信念却并无知识的盖梯尔反例仍然能够存在。我们马上就会看到这一点。尽管经典分析的捍卫者在努力弥补，但也有另外一些人已经由此受到启发，试图提出一种对知识本质的全新理解。

知识的因果理论

设想一下，你在即将结束一天的时候走进了隔壁的烘焙店找松饼，而你开心地发现展示橱窗里正摆着满满一筐松饼。“谢天谢地还没卖完！”你这样对自己说道。你所不知道的是，你看到的那个筐严格说来是装饰性的，那里面的“松饼”其实是塑料做的，只是为了摆出来吸引眼球之用，不管烘焙店里是否真的售卖松饼，它都日夜在那里摆放着。现在，幸运的事情来了，事实上烘焙店里的确还剩下了一些松饼，它们被摆在柜台下的架子上，没有让你直接看到。因此，你相信烘焙店里仍然有松饼，这个信念是真的，而且你也的确有理由相信这一点——假设那些塑料做的松饼模型足够逼真的话。但是，基于你所看到的展示橱窗里摆放的东西，你是否真的知道烘焙店今天的松饼仍然没有卖完？

50

在这个盖梯尔反例中，你的信念辩护的来源（那些塑料的装饰模型）与使你的信念为真的事实（藏在柜台下面的松饼）之间，存在着某个间隙。好的知识论一定要填平这个间隙。1967年，阿尔文·古德曼就提供了一个简单的新理论用于实现这个任务。按照他的“知识的因果理论”，经验知识要求认知者与某个事实之间具有恰当的因果关联。所以，按照这一理论，你对松饼的真判断之所以出了错，乃是因为它是由看到虚假模型而导致的。尽管烘焙店里的确有松饼，但这个事实并不是导致你相信“烘焙店里有松饼”的那个原因。另一方面，在经验知识的通常情况下，我们所相信的那个事实正是导致我们相信它的原因。例如，当你清晰地看到谷仓燃烧时，谷仓燃烧这一事实就导致你具有了某些视觉经验，进而使你形成了“谷仓在燃烧”的信念。记忆和证言也可以使我们建立与事实之间的因果关系，其中的因果链条会关联到我们所经历的先前事件，或是他人的经验。

方框3　知识的因果理论

S知道p，当且仅当事实p与“S相信p”之间具有恰当的因果关联。

（所谓“恰当的”知识生产的因果过程，包括记忆、知觉和推理。）

51

知识的经典分析主张我们的信念一定要是得到辩护的。明显地，这一要求在因果理论中消失了。古德曼的观点是，在很多情况下，即便人们并不能以证据证明他们的信念，这也并不妨碍他们获得知识。例如，通常受过教育的成年人都会知道，恺撒是被暗杀的，即便他们并不记得自己是从哪里得知这一事实的，也不能给出对这一主张的支持理由。但是，只要他们关于“恺撒被暗杀”的信念，与“恺撒被暗杀”这一事实之间，具有恰当的因果奠基关系，那么按照知识的因果理论，他们就是有知识的。在这里，所谓恰当的因果奠基关系，指的是从恺撒被暗杀的事实开始，经由对该事实目击者的知觉经验，再由目击者向古代的历史学家提供证言，历经几个世纪的史书记载传递，到达现在人们已经不记得名字的教师们的口中，由他们传授给受教育者关于恺撒被暗杀的信念。至于受教育者自己是否记得信念的准确来源，这一点并不重要；重要的是，他们的信念实际上有着恰当的因果链条，被锚定在所相信的那个事实之上。

因果理论被认为是知识论的一大进展，但它也还有诸多问题。有的批评者担心，如果不使用知识概念本身，我们如何能界定何为“恰当的”因果链条，而知识的概念恰恰又是我们试图去定义的东西。还有人指出，把知识与基于证据的辩护区分开是有问题的（第五章还能看到更多具体的困难）。其中，最深刻的一项忧虑恰恰来自古德曼自己，他在提出因果理论的十年之后指出了它的问题。事实证明，即便人们的确与事实之间具有恰当的因果关联，他们也依然有可能缺乏知识。

考虑下述反例：亨利一边带着他的小儿子在乡间开车兜风，一边辨别那些在身旁经过的地标。“看，儿子——那是一头牛！那里是一台拖拉机！那边的田野上有一个谷仓！”亨利相信田野上有一个谷仓，这是得自他对实际谷仓的知觉经验。按照因果理论的标准，亨利知道田野上有谷仓，（目前为止）因果理论似乎也符合我们的常识：我们通常说亨利知道那边有个谷仓，即便他并没有停下车，实际走过去看看再确认一下究竟是不是谷仓。谷仓足够大也足够显著了，你可以距离很远就能分辨出某个东西是谷仓。

然而，这里的例子又有一个转折。亨利并不知道，自己正行驶在“假谷仓县”的区域，古怪的当地人搭了几十个假谷仓，从高速公路的这一面看就跟真谷仓完全一样。纯粹是走运，亨利恰好在那时看到了这一片区域内唯一的一个真谷仓：如果他那时看到的是同一片区域内其他看上去相同的东西，他就会形成一个假信念，认为那里有个谷仓。那么亨利知道那里有谷仓吗？古德曼认为并没有，因为亨利出错的风险太高了。注意，亨利在做出“那边有谷仓”的判断时，实际上并没有形成任何假信

52

念，他对那个路边物体是谷仓的信念也是正当的和得到辩护的（假定他并不知道自己身处假谷仓县）。亨利的信念就其因果历史来说并无不妥，所以如果他没有真正的知识，那么就一定是知识的因果理论存在问题。

在这一点上，古德曼提出了一项新的知识分析，即把焦点放在可靠性而非因果性上：亨利的问题在于，他通常用于辨别谷仓的机制在假谷仓县中变得不可靠了。按照这个新的观点，知识就是由可靠的信念生产机制产生出来的真信念。这里的“可靠”意味着“可能产出真信念”。这种知识分析方法就是众所周知的可靠主义（reliabilism），它吸引了众多支持者。

可靠主义也引发了诸多批评。一种主要的批评意见着眼于作为其关键词的“可能”。如果知识必须是由很**可能**产生真信念的机制所产出的，那么成功的概率要有多高才是足够“可能”的？85%够吗？抑或是要99%？问题并不在于说有很多临界的例子，在其中难以辨别某人是否有知识。很多概念都有难以把握临界的例子。因此说，99%的可靠性对知识是足够的，而97%就不够，那么98%的情况就很难说到底还是不是知识了，这其实并不算什么特别让人烦恼的。真正的困难在于，从某一些因素的角度来看，我们似乎应该把那个标准提得特别高，而按照另一些因素，我们又得降低知识的可靠性标准。

为了看清楚把标准定高一点的好处，设想一下你有一张公平开奖的彩票，奖金是数额不菲的现金，中奖概率是千分之一。摇奖的环节已经结束，但还没有宣布谁是获奖者。就在你焦急等待中奖公告的时候，你是否已经知道你的这一张彩票没有中奖？它有99.9%的概率不会中奖，但是大多数人都会认为，尽管中奖的概率微乎其微，但直到中奖公告的那一刻之前，任何人都不知道自己不会中奖。但是，假如99.9%的概率都不足以高到算作知识，那到底多高的概率才能满足知识的要求呢？（你甚至还可以把这个概率再进一步地提高，只需设想你在参与的是一项更大规模的彩票活动，其间卖出了一百万张彩票，在那种情况下你也仍然不会仅凭概率上的微乎其微，就确切地**知道**自己不会中奖。）那么，你的信念形成机制是要高达99.999 9%的可靠性，才能算获得知识了吗？如果我们坚持认为是这样的话，那么看上去我们就正滑向怀疑论的深渊：在人们通过日常的感觉经验形成信念的过程中，即便有百万分之一的概率会产生假信念，我们也依然想要同意他们能够获得知识。就这一点来看，可靠主义的知识分析主张真信念一定要产生于生产真信念的可靠机制，虽然的确是有吸引力的简单方案，却不容易一下子看清楚究竟应该怎样解决这个问题。我们将在第五章中进一步讨论可靠主义的优劣细节。

山重水复疑无路？

盖梯尔指出，知识的经典分析定义错失了某些东西。哲学家们一开始希望很容易弥补上这块缺失的东西，只需要在经典定义的三个条件之外再增加第四个条件，或是用全新的方式来把知识分解成更简单的单元要素。这些希望最终都落空了。在盖梯尔之后的几十年里，人们提出了数十种知识的分析定义，但没有一种能够得到广泛的支持。许多新的知识分析也很容易受到新的直觉反例的挑战，或是陷入某种恶性循环，把本来应该试图去定义的概念当成了定义中预设了的东西，而且往往是被新引入的某些技术性词项掩盖了。经历了三十年不断复杂化却令人失望的探索之后，有的哲学家已经开始怀疑知识的分析根本就是一个无解的问题。但为什么它无解呢？

答案或许隐含在辩护与成真的关系之中。通常我们认为，假信念是可能得到辩护的。辩护与知识不同，它并不保证被辩护为正当的东西就一定为真。假如一个人看到了非常逼真的松饼的全息影像，那么相信在他面前有一块松饼，对他来说并不算是不可理喻的事情，反倒是很合情合理的，即便他的信念是假的或只是碰巧为真的。假如一个侦查员碰到了大量误导性的证据，而这些最终使他产生了假信念，我们也可以说他是完全合情合理地拥有假信念。譬如，基于他所发现的东西，他可以完全合乎情理地相信某个管家有罪。实际上，正是由于可能存在着得到辩护的假信念，知识的经典分析定义才遭遇了其最原初的沉重打击：我们要问何种正当信念能算作知识，那么成真条件就把范围缩小到“仅仅是那些为真的信念”。但是，一旦我们允许理由的辩护并不总是导向信念的成真，那么盖梯尔的问题看起来就无可逃避了。由于美国哲学家琳达·扎克泽博斯基的贡献，现在甚至都有了构造知识定义的反例的标准配方，具体说来是这样做的：先描述一个得到辩护的假信念的情况——或是更一般地说，是针对某个知识定义的要求，满足除了成真条件以外的

55

所有条件的假信念。然后，加上一个幸运的转折，以便使被相信的那个命题最终为真。扎克泽博斯基的这个配方就能为众多知识论观点构造大量反例。

“知道”究竟能不能被分析？

如果知识分析的工作进展不顺利，那么不妨退回来想想究

竟是为什么。有的哲学家就提出，真正的问题在于**知识**本就不是一个循规蹈矩的概念：例如，马特·维纳就认为，我们对“知道”这个动词的用法就有一些虽然顺手但并不严谨一致的原则。我们在直觉上使用这个动词的时候有很多相互冲突的方式，要想从中得出一个对知识的清晰定义，就好比是要鉴别出一台由各色零件组成的汽车的品牌和型号。还有的哲学家主张，我们应该把注意力从知识的概念转向其他话题，譬如什么是合乎情理的信念等问题。但是，即便我们在分析知识的过程中经历了诸多困难，这也并不能构成使我们离开这个任务的理由。有时，分析定义中碰到的阻力是好事情，它意味着我们触碰到了某种更为根本的东西。

为了更清楚地体现这种阻力的价值，可以先来看一下阻力较小时的情形。平常用的食盐很容易被分析为氯化钠（$NaCl$），其中的组分（Na^+和Cl^-）都比它们组成的化合物要简单得多，而且也可以与其他材料形成不同的化合物。那么，知识分析的任务从一开始也持有类似的观念，即知识也是由更简单的物质所构成，其中就包括真与信念。当然，真和信念也可以分别与其他材料结合形成不同于知识的状态——例如，当信念与为假的因素结合，就不能称之为知识。但是，在知识分析的任务中，另一个重要的前提假设却是更成问题的：作为知识组分的信念与真难道真的比知识本身简单得多吗？知识真的是由这些要素（甚或还有其他要素）构造出来的复合物吗？

56

或许并不是。具体说来，**相信**并不比**知道**更简单，也并不比它更基本。如果说**知道**是基础性观念而**相信**只是它的派生物，这样行不行呢？这种观念就是当代知识论中被称作“知识优先”（knowledge-first）的思潮。作为这一思潮的领军人物，蒂莫西·威廉姆森论证说，哲学家们之所以没能得到满意的知识定义，乃是因为他们试图用信念加上其他东西来分析知识，而实际上知识的概念反倒比信念的概念要更为基础和本源。

知识优先的进路初看起来有一些违反直觉的方面。似乎信念不得不被当作构成知识更基本的组分，有一个很简单的理由是信念状态要远远多于知识。每当人们获知某些事情时，我们都可以说他们相信这些事，但反过来则不然：例如，人们会有一些虚假的或非理性的信念，这不能算作知识。回到我们原初的那个类比，基础组分在很多情况下都要比它们所组成的东西更常见。譬如，钠离子要比食盐更常见，因为它并不只存在于氯化钠中，还存在于诸如硝酸钠等其他化合物中。你有食盐的时候就会有钠离子，但反过来却不行。然而，更常见的东西却未必总是更基本。考虑一下完美的圆的概念：它非常简单，在自然界中也很难找到。现在，如果“圆形”意味着“至少大致是圆的”，那么显然圆形的东西就远远多于圆的，但圆的概念却依然是更为基本的。圆的概念是我们的根本出发点，是我们用来定义“圆形”的东西。我们有理由相信，圆的几何本质整洁而简单，但近似圆的几何性质却是复杂而凌乱。

57

知识与信念之间的关系，在很多方面都类似于圆与圆形的关系：知识是理想性的，而信念则是某种接近于此理想的东西。根据“知识优先”的知识论观点，试图用信念加上其他要素来定义知识，乃是非常糟糕的想法。正像不可能用一个近似的圆形加上其他要素来定义圆的概念本身一样。按照威廉姆森的想法，我们不应该把知识解释为信念与其他要素的复合状态，而是

方框4　古代盖梯尔反例

在西方哲学中，1963年被当作一个特殊的年份，正是在这一年发现了盖梯尔提出的那些反例，表明了在得到辩护的真信念与知识之间的间隙。但实际上，早在大约公元770年的一篇文献中，印度哲学家法上已经提出了如下例子：

点起一堆火来烤肉，那火尚未冒出任何烟来，但肉的香味已经引来大批昆虫。隔开一段距离以外，一位观察者看到黑压压的一群飞虫出现在地平线上方，而误把它当成了烟。于是，这位远距离的观察者说："那里正在烧火。"

➢　观察者是否真的**知道**在远处烧着一堆火？

沙漠中的旅行者正在找水。他看到前方的山谷中有一片闪着蓝光的区域。不幸的是，那只不过是海市蜃楼。但幸运的是，当他走到那片看上去有水的地方时，那里实际上真的有水隐藏在岩石之下。

➢　旅行者在产生幻觉的时候是否真的**知道**前面那块地方有水？

这些例子都包含了基于某些好证据的真信念。法上与盖梯尔一样，也是把这些反例用于反驳对立的知识论观点。类似的很多例子也都在追随法上的其他印藏知识论学者那里得到积极的讨论。甚至有些1963年以后在西方哲学中提出的应对这些反例的方案，早在几个世纪之前的印藏知识论传统里已出现过了。例如，14世纪的甘吉沙已经发展出了一种较为详尽的知识因果理论。

应该反过来用“知道”来解释“相信”：“粗略地说，相信p意味着就像知道p那样对待p。”在他看来，知道是一种本质上要求是正确的心智状态，而相信则是以正确性为其理想目标的心智状态，而它可能会达不到那个目标。

在当代知识论中，有太多都是从真信念加上其他要素的角度来分析知识，所以宣称知识比信念更为基本是有些革命性。假如知识的分析定义是知识论者所能做的唯一事情，那么宣称知识更为根本乃至于拒绝被分析，就意味着整个知识论工作都要停下来。然而，即便知识不是信念与其他要素的复合物，也还有很多值得研究的课题。譬如，我们可以研究知识与辩护的关系，研究知觉与证言如何产生并传递知识，研究知识在不同视角下的形象显现。这些议题不论是在“知识优先”还是在“信念优先”的知识论中都是重要的。其中有一个问题对于这两方面（甚至对于两者之间的争论）都特别重要，那就是从第一人称向第三人称视角的转换所发生的差异。下一章将会详细考察这种转换的意义。

59

第五章

内在论与外在论

第一人称观点

在阅读这一章之前，你大概已经认为自己知道如下事实：珠穆朗玛峰是世界的最高峰。但是，如果这个关于珠穆朗玛峰的事实不是你刚刚得知的，你或许并不知道自己最初究竟是从哪里得知这一事实，或是其他任何随机琐碎事实的。根据研究记忆的心理学家的意见，除非是你在听到有关珠穆朗玛峰的事实时发生了某些关键的生活经验，譬如在你的小学地理课堂上提到世界最高峰时发生了地震，否则你根本不会记得是从哪里听来的。事实上，如果有人质疑你是不是真的知道“珠穆朗玛峰是世界的最高峰”，你或许也没有太多能为之辩白的东西。你也许可以说它似乎是一个熟悉的事实，但质疑者也可以反驳说这种熟悉的感觉很可能是欺骗性的。对世界上很多人来说，“悉尼是澳大利亚的首都”听起来也像是个广为人知的事实。你在感觉上非常确定自己是对的，有时却恰恰实际上是错的。

假设一个人主张珠穆朗玛峰是世界的最高峰，但并未意识60到任何能够证明这一主张的东西，那么他是否仍然有真正的知识？在这个问题上，哲学家们划分成两大阵营。内在论的一方说：假如你真的没想到任何支撑性证据，那就麻烦了。因为，如果你无法获得任何支持这一主张的东西，那么你关于珠穆朗玛峰的信念就不能算作真正的知识。与之相对比的是你的另一些信念。譬如说，你相信自己正在读书，你对这个信念就可以给出证明，只需诉诸那些你意识到的当下具有的经验。它也不像你的另一些信念，譬如说你相信没有最大的素数，你可以自己重演一遍欧几里得证明的步骤来证明它——希望你还记得那些证明步骤。一言以蔽之，知识的基础和根据就在于你自己的经验与推理能力。内在论者特别强调的是你利用第一人称视角中的资源所能做的事情：如果你不能自己理解为什么应该如此这般相信，你实际上就没有知识。知识之所以区别于瞎猜等更低的状态，其中最根本的一点就在于认识的主体自己能觉察到知识的良好根据。

相反地，在对立一方的外在论阵营看来，知识是个人与事实之间的某种关系，而即便个人并不在内在论的意义上满足以第一人称获得根据的要求，这种关联性仍然可以存在。如果珠穆朗玛峰事实上的确是世界的最高峰，如果你也以某种恰当的方式关联到了这一事实——稍后我们再详细解释什么是“恰当的方式”——那么，即便你不能解释何以自己会认为“珠穆朗玛峰是世界的最高峰”，你也仍然可以具有这一知识。外在论者可以毫无惧色地承认，**在有些情况下**，你的确不仅知道某件事情，而且也对你如何知道这一点拥有独特的第一人称洞察。但是，从外在论的角度说，对如何知道的洞见是一项选择性的额外好处，并不是在每一项知识的实例中都普遍具有的东西。外在论者论证说，如果我们**总是**要求对如何知道这一点保持第一人称的洞察，那么我们就会面临陷入论证上恶性后退的风险。因为，按照内在论的思路，你对于自己如何知道这一点不应该仅有一些随机的想法，而应该真的**知道**你如何知道这一点——如果没有知识，洞见又是什么呢？但如果知识总是要求“知道你如何知道这一点”，那么这第二层的知识又要求它自己的内在论保证，也就是“知道你如何知道你有这点知识的”。如此便无穷后退下去。于是，对于一件最简单的事实，你想要知道它也得有无穷多层的洞见才行。所以外在论者认为，内在论的思路有把我们带向怀疑论的危险。 61

那么，第一人称视角究竟在知识中扮演怎样的角色？它是使知识区别于单纯的信念不可或缺的组成部分吗？抑或是这样的主张就会引向怀疑论？尽管远至18世纪就已经有内在论与外在论的知识论对立，直接的争论却是在盖梯尔对经典知识定义提出挑战后才出现的。因为盖梯尔的挑战提出了在知识证成方面的严重困难，进而又推动了外在论的知识理论的发展，例如第四章讨论到的古德曼的早期因果理论与晚近的可靠主义理论。而为了更广义地了解外在论的逻辑理路，我们有必要考察一种更有影响力的外在论观点——罗伯特·诺齐克提出的知识的追溯理论。

诺齐克的追溯理论

诺齐克的外在论来自这样一个基本观念：有知识的人不仅有某个给定问题的正确答案，而且假设答案已经变更了，他也还是**会**正确地回答那个问题。设想一下，有个医生告诉你你感染了甲肝。如果这个医生对他所有的病人都这么说，而且在大多数时候都是错的，那么即便他这次是对的，你的确得了甲肝，他也并不真的拥有关于你病情的知识。而要真的具备知识，你的医生就必须能对感染了的病患做出阳性的诊断，而不会对那些根本没病的人做出错误判断。所以，诺齐克认为，所谓知识不过是这样一种东西：如果你总是倾向于相信那些为真的东西，而不相信那些为假的东西，那么你就有知识了。注意，诺齐克的理论并没有特别要求觉察到自己的理由或根据。还是以那个医生为例，假如他就是对甲肝有特别可靠的诊断直觉，即便他也不能解释自己究竟是从病人的什么特征上判断出来的，只要他的判断正确，那么他就是有知识，知道你是不是罹患疾病。甚至即便他对自己做判断的过程有**错误**的认知，他也还是可以算作有知识。假设他自以为所依靠的是检验报告，而实际上他只是基于病患的肤色与气味上的微妙线索就做出了诊断，那么在追溯理论看来，只要他的诊断的确追溯到了是否罹患疾病的事实，那么他就仍然具备对病患病情的知识。

62

方框5　知识的追溯理论

S拥有知识p，当且仅当：

（1）p为真；

（2）S相信p；

（3）假如p不为真，S就不会相信p；

（4）假如p为真，S就会相信p。

追溯理论的基本结构包含四个条件：条件（1）和（2）是有关实际上发生的事情。条件（3）和（4）是虚拟条件句，说的是在不同于现实的情境中**会**发生什么事情。根据条件（3），你相信自己现在正在读书，这一信念应该以这样一种方式形成，即对你不在读书的那些情况保持敏感：这也就是说，假如你事实上没在读书，你就**不会相信你在读书**。一个对读书总是有错觉的人，即便他真的在读书，也并不知道自己在读。同时，按照条件（4），你关于自己正在读书的信念还必须以对真实的读书活动保持警觉的方式形成。你这次知道自己在干什么，不应该只是一次侥幸，你必须对自己是否在读书这件事有一种可靠的见证，才能真正知道自己现在在做什么。（3）和（4）就是知识的"追溯"条件：真信念要想成为知识，就必须以某种方式追溯事实。那些能够成为知识的信念必须能够随着你周围世界的变化而变化；它们不仅要符合当下的世界，而且，只要事情稍微变得有点不一样，它们也能随时呼应世界会变成的那个样子。如果你有知识，你就应该安全地锚定在通往真的道路上。

追溯理论也有它自身的问题。其中一个问题与我们获得知识的方式有关。诺齐克自己意识到，在有些情况下，信念之为真与我们形成该信念的方法纠结在一起，难以处理。比如下面这个例子。假设一位老奶奶躺在医院里，她的宝贝孙子来看她。当他站到病床前，奶奶看着他，于是相信他身体不错，至少足够好到来看她。然而，我们假设家里人都很关心奶奶，不会让她知道一点孙子身体不好的信息。即便孙子身体不好，家人也还是会告诉奶奶说他很好，以免让她担心。所以，假如奶奶所相信的其实不是事实，也就是说她的孙子其实身体并不好，那么奶奶也不会比现在更明智：她也还是会继续相信孙子身体很好，这就违背了追溯理论的条件（3）。但是，我们毕竟还是会觉得，只要孙子的确身体好，且又来到了病床前，那么奶奶就会知道孙子身体足够好，能来看她。这种知识并不会受到家人的秘密计划影响，那个计划即当孙子生病时应该怎么做。所以，为了解决这个问题，诺齐克对追溯理论提出一项修正，使之囊括了对“信念形成方法”的考量。在上述例子中，奶奶的信念形成方法并没有什么问题，她就是通过看着来到自己床前的孙子而获得信念；假如孙子生病了，这并不会使她的信念形成方法出问题，因而不能导向真理，而是她不能再用这个方法了。那时她可能就会用其他方法来形成信念了，譬如相信关心她但是会欺骗她的家人的证言。

与古德曼的可靠主义一样，诺齐克的追溯理论最后也非常依赖于信念形成方法的概念。当信念的形成方法满足某些标准，且你又用这一方法找到真理时，你就拥有知识。按照外在论观点，你不必自己知道那些标准是什么，你甚至都不用知道自己用的是什么方法来形成信念。但是，不管你用什么方法或机制形成了信念，它们必须实际上追溯到了事实（根据诺齐克的追溯理论），或是可靠地以较高的比例产出了真信念（根据古德曼的可靠主义）。但是，如何决定认知者实际上采用的是什么信念形成方法？外在论在这个问题上面临很大的困难，这个难题实际上古德曼在刚提出可靠主义的时候就想到了，后来又被他的内在论批评者们不断地加以强调。这个困难如今被称为“普遍性难题”。

普遍性难题

还记得第四章中讨论的亨利开车经过假谷仓县的案例吗？古德曼论证说，尽管亨利相信他看到的是谷仓，而且这个信念也的确是由恰当的东西导致的，即他所看到的事实上就是谷仓，但是亨利也依然不知道他所看到的是谷仓。他只是很幸运地在一堆假谷仓中看到了一个真的而已。古德曼的分析结论是，在假谷仓县这个环境中，亨利用来辨别谷仓状物体的机制乃是不可靠的。在假谷仓县，这个认知机制基本上总是产出假信念，所以即便它这次提供给亨利的是真理，亨利也不能算有知识，因为知识要求有可靠的机制。这个结论之所以有道理，是因为我们首先要赞同古德曼的下述前提：其一是所考察的环境仅限于假谷仓县，不能扩展到以整个世界为论域；其二则是亨利所用的认知机制是他专门用来辨别谷仓的机制。假如我们现在讨论的是亨利依赖于某个更一般的机制，比如“视觉”，那么上述可靠主义分析的结论就不成立了，因为视觉机制即便在假谷仓县也依然是可靠的。（亨利通过视觉来识别路上画的交通标志线、路边的树等等，而且他的确看到了所有这些东西——假谷仓县完全没有抹杀视觉作为一个整体的可靠性。）因此，如果要得出“亨利的信念形成不可靠”这个结论，我们就必须更具体地界定他所采用的认知机制与所处的环境。但如果我们再具体一些，为什么不能直接说这一机制就是“他在现场看到那些谷仓状的东西时，采用的谷仓状物体辨别力”？既然亨利现在看到的是一个真谷仓，那么这个最具体的机制总是会给亨利提供真信念。但如果我们如此狭义地界定信念形成的过程，那么任何真信念就都可以被当作知识。所以，这里的问题是，我们如何以恰当的细节层次来切中描述机制及其环境的目标？

外在论者提出了几种针对普遍性问题的可能应对路径。古德曼在1976年提到的（如果不是赞同的话）一种途径是，承认在描述亨利所依赖的机制这个问题上，并不存在任何简单的客观事实。因为对亨利是否有知识这个问题而言，也不存在任何客观的事实。如果我们按古德曼最初描述他的方式，那么亨利就没有知识；但如果我们采取更为宽广的视角，或许他就有知识了。在第七章中我们会单独讨论这样一些理论，因为它们很复杂，允许有对知识状态的多元断定。外在论者采取的另一种路径是主张的确有一种划定边界的自然方式，以便确定在任何给定实例中得以形成信念的过程。他们也提供了多种不同的方案，借用了从自然语言的各种模式到信念形成的科学等种种理论。外在论者把这个问题视作一个重要的、有发展前途的研究领域，而内在论者却对能否找到满意的答案持怀疑的态度。

同时，还有的外在论者主张，普遍性问题并非外在论者的专属问题：内在论者在某些方面可能同样避免不了。内在论的理论不会明显地聚焦于信念形成的方法：他们更多地讨论人们所能得到的证据，以便证明或支持信念。但是，即便是内在论者也需要关心实际形成信念的方式，因为下述两者间的差异对所有知识论学者而言都是重要的：一面是有很好的理由去相信某事，另一面则是在那些好理由的基础上实际地具有信念。设想有这样一个法官，他有关于被告罪行的充足证据，而且也相信被告有罪，但这一信念却是源自他的种族偏见而非基于证据。那么对知识论学者而言，不管是持何种学术观点，都会想要批评该法官的信念形成途径，视之为缺乏根据的。但是，只要我们关心的是信念产生的实际过程——而不只是认知者所能够获得的证据——那么我们就需要对这些过程给出解释，同样需要找到某些描述它们的方式，使其既不过于严格也不过于宽泛。在这个意义上，普遍性问题似乎是所有人都必须面对的考验，只不过它在外在论者那里表现得更为显著和迫切一些。

可质疑的方法

内在论区别于外在论的地方就在于，它给认知主体的视角以某种特殊地位。内在论者主张，知识要求认知主体能从自己的反思或某种直接意识中看到某些东西。内在论的不同观点之间的分歧，也就在于要求认知主体看到什么，是要有某些证明性的证据，还是全部的证据，抑或是否要求认知主体知道自己的证明。但核心的观点都是认知主体必须理性地思考，基于他所能获取的证据来做出决定。外在论者则认为，即便认知主体无法获得证据，也依然有可能具备知识。他们主张，这一举动使他们能够捍卫许多我们通常想要归类为知识的信念，例如你关于“珠穆朗玛峰是世界的最高峰”的信念。在这个例子上，内在论者可以有两种立场。一种是主张它不应该算作知识，至少对于我们大多数普通人来说都是这样，除了山地地理学方面的专业人士，他们或许真的拥有这项知识。持这种观点的内在论者会说，通常之所以说普通人“知道”珠穆朗玛峰是最高峰，主要是在宽泛的意义上使用这个概念了。就好比我们也会说，法国领土的轮廓就像一个六边形——这对于日常的需求而言足够用了，尽管它实际上并不真的是六边形。另一些内在论者则认为，我们事实上已经能够内在地获得对这类信念的辩护证据了：人们可以意识到自己的记忆力很好，而且又记得自己多次听过或者读到过这个判断，诸如此类。尽管记忆给出的有些判断是错的，例如“悉尼是澳大利亚的首都”，但内在论者只需要要求认知主体能够获得证明性的证据，而大多数并不需要进一步要求认知主体具备能完全保证信念为真的证据。

内在论者无须要求认知主体具备“知道自己有知识”的特殊能力，或是能够获得保证所知为真的证据：温和一些的内在论立场仅仅要求认知主体应该具备获取证明性证据的能力。美国知识论学者劳伦斯·邦茹曾经用下面这个例子来说明为什么要求具备知识的证据。设想一下，有个叫萨曼莎的人认为自己在某些特殊主题上具备千里眼的能力。萨曼莎认为她的超自然能力能使其“看到”任意时刻美国总统的位置，尽管她并没有很好的理由相信她具备此种能力。她甚至没有检验过她这种超自然的“看”是否有准确性，也并没有很好的理由相信千里眼能力真的存在。有一天她借助这种能力相信美国总统在纽约，尽管她在报纸上读到总统今天在白宫，而且从电视直播也可以看到总统正在华盛顿特区开记者招待会。尽管如此，萨曼莎仍然拒斥了所有这些普通的证据，而相信她那种超自然的视觉所看到的东西。那么萨曼莎是否真的知道总统此时在纽约呢？

在回答这个问题之前，先让我们来看一下这个故事包含的几个关键转折。首先，总统此时的确是在纽约，所有那些证明总统在华盛顿的“普通证据”，实际上都是出于一项国家安全危机的考虑而迷惑公众的把戏。此外，尽管萨曼莎并无好的理由相信其特异功能，她实际上想对了，也就是说，不管是如何做到的，她对总统的位置**的确**有某种超自然的可靠把握。那么，邦茹的结论是，如果按照外在论的标准，萨曼莎就的确知道总统此时在纽约，因为她的信念为真且是可靠的。如果她的信念的确是由某个可靠的机制形成的，那么这就不是碰运气的猜想。然而，萨曼莎的确没有理由拒斥掉那些与她的判断相反的普通证据，而且在她并不知道自己的千里眼能力是否可靠的前提下，她就选择追随超自然能力给出的结论，这未免也是不合情理的。至少我们可以这样说，如果一项知识理论主张某人拥有对某个事实的知识，同时却又承认他在这一事实上做出了不合理的选择，那么这个理论的确难以自圆其说。

针对邦茹的这个批评，外在论者发展了好多不同的回应。一种回应承认，说萨曼莎有知识似乎会有些诡异，但又坚持她实际上的确就是有知识的。这种诡异的感觉来自人们对超自然能力本身的不熟悉，或是在我们的文化中根本就不相信那些声称有这类能力的人。外在论者指出，当我们基于日常的感官信息形成信念，事实上我们每个人就进入萨曼莎那样的处境：我们并没有很好的理由相信视觉的可靠性，除非是我们已经把视觉当作信息的可靠来源。但是，在萨曼莎拒斥相反证据的问题上，这一回应依然有不能完全解释的地方。另一种外在论的回应则突出强调了这一点，既然邦茹主张人们可以可靠地获取信念却没有知识，那么外在论的回应就要证明这种主张何以不成立。如果萨曼莎故意忽略反面证据，那么她就是在以一种通常并不产生真信念的方式思考，因为故意忽略证据的人通常也就不会可靠地得到真信念。在这个意义上，萨曼莎的信念形成过程的确开始于某个可靠的部分，即对总统位置的超自然视觉输入——

但是，既然她接着又是以忽略证据的方式，也就是以不可靠的方式思考，那么这整个的信念形成机制也就完全不可靠了。如果这整个的认识方法并没有很好地符合外在论的标准，那么外在论者当然也就可以解释为什么萨曼莎没有知识了。自然地，这种对邦茹挑战的回应又把我们带回了普遍性难题上的困境：我们到底该如何确定哪种认知机制的描述是恰当合理的？

另一种不那么对立性的回应会承认内在论也有某些合理性，同时通过添加一个额外的条件来软化外在论的立场。这个额外的条件就是，外在论不仅要求能可靠地追溯到真理，而且还要求认知主体不能有任何可获取的且无法合理忽视掉的反面证据。内在论者依然反对这种折中的努力：既然你都不要求能够从第一人称的方式获得正面支持性的证据，何以还要关心能否获得第一人称的反面证据呢？还有一种折中的立场是这样的：它承认证据之可获得与否对于某些信念的形成过程来说依然重要，但又坚持这不是因为知识普遍要求第一人称条件。当代心理学已经发现人们有两种不同的思维方式。一种是迅速的、自动的思维，它不需要我们意识到任何思维步骤就能直接给出答案。譬如说，7乘以11等于多少？这答案一下子就想到了。这种思维方式在帮助我们识别出熟悉的人或陈述熟悉的事方面特别有用。另一种是慢速的、受控的思维，它需要在几个步骤组成的序列中实现，譬如说现在转而问你7除以11是多少，你要得到答案就得靠这种思维方式了。它可以帮我们解答数独或是计算应缴的税款。对于这第二种思维方式来说，人们必须能够以第一人称的方式获取所有步骤。许多问题都可以用这两种思维方式之一来处理。例如，别人问你昨晚吃的什么，你可能就直接回

70

答了出来，想到什么就说什么，而且还很有可能是对的。又或者你要仔细地想一会儿，努力回想一下昨天晚饭时发生的事情，重新激活吃饭过程中的细节。我们通常默认会采用自动的思维，但当我们需要意识到自我，或是预见到需要为自己的答案辩护时，我们就会切换到更具系统性的思维方式上。对知识所做的系统反思恰恰需要这种思维，它也是我们希望萨曼莎在面对反面证据时所应该采取的思维方式。在第一人称意义上获取证据的能力对这种系统思维方式很关键，这一点也许让内在论者感到欣慰，但外在论者也同样会满意，因为自动的思维方式也被广泛认可为知识的来源，比如当你认出来一位朋友的时候。外在论者还可以论证说，之所以第一人称的证据获取对系统性思维很重要，也不过是因为它从根本上保证了这一思维方式能够可靠地运作起来。

我们越是深入地理解第一人称视角所能获得的东西，我们就越是有可能进一步理解这一视角对知识获取的真正意义。同时，检验内在论与外在论的众多对立的知识理论，也很有可能取得进步。接下来要特别考察的也是特别有用的知识类型，就是从证言中获得的知识。

第六章

证言知识

别人是这么说的

在知识的领域中，我们许多珍贵的存货都是二手的。我们依赖于他人才得以掌握很多事情，从遥远地域的地理情况到朋友生活中的平凡故事，莫不如此。如果我们不能利用他人这个资源，就会丧失对很多话题的理解，这既包括古代的历史——除了我们可以通过个人的考古探险来发现的以外，也包括名人的婚礼——除非是我们也在受邀之列。显然，证言扩大了我们的视野，但问题在于解释这一切是如何做到的，以及能扩大到什么程度。听别人的话或者读别人写的东西，是否会以某种独一无二的方式给予我们知识？为了从别人那里获得知识，我们是否需要特殊的理由来信任他人？对于像维基百科这样的信息来源，其中绝大多数文章皆出自多人或匿名作者之手，我们又该如何看待？

一个极端的观点是，正如有些哲学家所主张的，证言**从没有**

真正提供知识，约翰·洛克就是这一立场的著名代表。而在另一个极端，也有的哲学家论证说，证言不仅提供知识，而且还是以独特的方式提供知识的。按照这一观点，证言就是获取知识的特殊途径，这一路径与感觉经验和推理同样根本和重要。我 72 们可以在古代印度哲学中找到这种观点，而它在当代英美哲学中也非常盛行。通过考察这两个极端的观点，以及一些主要的居中立场，我们就可能找出某些要素，它们被广泛认为对吸收他人的证言起到了非常重要的作用。

求知无门

证言什么时候能提供知识？有的哲学家认为，“从来没有过”。那么，为什么这些哲学家会怀疑证言知识的存在，即便并不同样地怀疑其他类型的知识呢？那就首先要厘清什么是所谓的“证言”。别人在证言的行为中告诉你某些事情——这既可以是通过言谈举止，也可以是通过写下来的东西——总之，他们告诉你的内容就是你在这项交流中的最大收获。即便是那些怀疑证言知识的人，也可能同意，倾听或阅读别人的言辞会产生普通的知觉知识。例如，设想一下，你看到有人在纸片上写下这句话：“我的书法很漂亮。”或者你听到别人说：“我的声音沙哑了。”如果你的确看到写下来的字很漂亮，或是听到了沙哑的声音，那么你也就知道别人的所写与所说乃是真的。但是，你此刻的知识是知觉性的，而非证言性的，因为所写所说的内容并没有在你的获知过程中产生特殊的作用：如果写下或说出的句子是“史密斯得到了那份工作”，那么它可以同样起到展现书法之美与声音之哑的作用。如果你基于我的证言相信了某件事情，那么你就理解我所说的话,并且也相信它。

在洛克看来,知觉知识与我们从证言中获得的东西之间有着明确的对立。例如,前者是知道你正听到的那个声音是沙哑的,后者是得到消息说史密斯得到了那份工作。关键的差异在于确定性,而洛克认为这是知识的必要条件。因为知觉可以给你对某件事情的直接的确定性,正如你直觉上就知道红色不是黑色那样,你就由此得到了知觉知识。洛克说,我们从证言中得到的东西最多也不过是有很高的可能性,而不会是有确定性的东西。在英格兰的冬日里,如果你看到一个人在冰封的湖面上行走,你就知道那个人正穿越这片湖。如果是别人告诉你说他看见有人正走过湖面,那么,只要你的信息源值得信任,且他所说的事情也符合你过往的观察,那么你就有理由相信这个报告很有可能为真,但你实际上对此仍然没有真正的知识。

你自己的背景经验发挥了重要的作用。洛克讲了这样一个故事:暹罗[①]国王听荷兰大使说起,在荷兰的冬季水会变得足够牢固,能撑起一个成年人甚至是一头大象的重量(前提是你真能引诱一头大象冬天去荷兰的话)。据说,那位国王回答说:“迄今为止你所告诉我的所有诡异事情我都相信,因为我看你是个足够清醒稳重的人,但现在我敢肯定你是说谎了。”洛克对于这位持怀疑态度的国王深表同情,因为他所有过往的经验都来自热带地区,对他而言,比起相信水会自然地变成固态,相信大使很有可能在说谎才更合理一些。

即便证言不可能很有确定性,它也或多或少可能为真,而洛

① 今泰国。——编注

克的建议是，从你已有的证据效力上来决定对证言持有多大的信心。他用一个复杂的公式来决定证言中合理的信心程度。首先，你要权衡一下它在多大程度上符合你的过往经验，然后你要考虑如下六个方面的因素：

1. 有多少证人；

2. 证人的诚实性如何；

3. 证人的技能水平；

4. 证人提供报告的目的；

5. 报告内容的内在一致性，以及你听到这一报告时的具体情境；

6. 有没有与之相反的证言。

74

尽管洛克认为证言并不传递知识，他也不认为我们应该笼统地拒绝相信别人告诉我们的事情。洛克说，一个通情达理的人一般会同意别人的证言。如果证言符合我们的经验，而且在上述六点检查清单中获得较高的分数，那么我们就从证言中获得了在实践目的上非常像知识的东西——“我们很容易接受它，且牢固地立足于它做出发展，仿佛它是确定的知识那样”。但是，我们从证言中得到的东西依然不同于“红色不是黑色”这样的发现。按照洛克的观点，证言中得到的东西仍然有可能在未来的经验中被颠覆。例如，你曾认为自己没有任何理由不信任他，而他告诉你看到有人横穿湖面是他虚构的，后来你又听十多个人说，那里的冰今天太薄了，根本无法在上面行走。洛克认为，这种对未来反面报告的脆弱性表明，我们从证言中得到的东西并不真的等同于知识。

稍后我们会简单分析一下对洛克论证的挑战，但在此之前，有必要先看一下他的观点实际上是多么激进。如果洛克是对的，那么当有人问“你知道你在哪儿出生的？”时，最合适的答案是“不知道”（假定与其他大多数人一样，你在这个问题上的信念是你的家人告诉你的，或是写在你的出生证明上的）。你可以说你很有可能出生于某地，但你对此从没有第一手的经验，因此你并不真的拥有知识。你也并不知道乔治·华盛顿曾是美国总统，或是地球上真的有南极洲（假设你从来没去过南极）。如果本书的读者是一个洛克主义者，他们甚至不能说自己知道洛克曾经在这个世界上活过，他们最多只能认为洛克很有可能活过。

显然，洛克会反对我们通常的一些说法，因为我们会很随意地把人们描述成从证言中获得知识。例如，“琼斯知道史密斯得到那份工作了吗？”——“是的，他已经知道了。我刚刚告诉他的。”然而，我们也会承认，在很多情况下这些通常的说法并不严格准确，就像我们说太阳升起来或者落下去，但实际上是地球在旋转那样。那么洛克是否有很好的理由认为，我们严格说来并没有从证言中获得知识？从对未来怀疑的脆弱性上论证证言不能给出知识，这是有问题的，部分是因为同样的理由也可以适用于由知觉和记忆给出的判断，而这些恰恰是洛克想要认可为知识的东西。洛克认为，知觉能给我们以知识，例如你现在在读书，而且过会儿你也会记得现在在读书，这就是从记忆中保持住了你的知觉知识。然而，你也有可能过后会怀疑你自己。即便你现在的确感知到了某些东西，你也可能将来怀疑它们，比如怀疑自己是否只是在做梦。但只要你当下的确是在感知而非做梦，洛克就不认为这种将来的怀疑会剥夺你当下主张的知识。但是，类似的论证却适用于证言：如果有人告诉你，史密斯得到了那份工作，而你实际上也没有对所说内容产生怀疑，那么你就应该获得了知识所要求的那种确定性。如果你过后又开始怀疑了，譬如说获得了某些相反的证言，那么你就可能失去那项知识，但这并不能证明你从来没有获得过那项知识。如果你的信息源是有知识的，那么你随后的怀疑就不能证明最初的判断非真：如果你的信息源知道史密斯得到了那份工作，那么这一点就必须为真。任何与之相反的报告都是迷惑性的。当然，在某些情况下你没有产生怀疑，直接采纳了说谎者的说法，把他说的当成真理，而这样的情况就类似于知觉中发生错觉时的情形。如果说在提供知识的能力方面，知觉与证言之间存在很大的差异，那么洛克并没有告诉我们这种差异究竟是什么。

中间地带：还原论

尽管洛克主张证言不能提供知识，但这种观点属于少数派。大多数哲学家都持有更为积极的态度。还原论就是一种较为温和的主流积极态度：我们的确从证言中获得知识，但证言提供知识的能力并没有什么特殊的。我们在阅读文字、倾听言谈或是观察他人的手势与符号语言时，都是在从普通的感觉经验中获得证言。假设一切都进行得十分顺利，感性知觉就会让我们知道说话者说出了某个句子。为了从这个句子中获得所说内容的知识，而不仅仅是“说话者如此这般地说”这个事实，我们就通常需要依赖推理与知觉的能力。这里仍然存留有某种洛克主义的东西：你去检查洛克开出的那个清单，以便考察证言为真的可能性——例如它与过往经验是否符合，说话者的诚实性是否有证据可以证明，等等。但是，当你所听到的证言在这个清单中得分足够高时，你就真的获得了关于那个在交流中呈现的命题的知识。这种思考证言的方式之所以被称为还原论，是因为它主张证言提供知识的能力可以还原为其他信息源提供知识的能力，特别是知觉、记忆和推理。

还原论有两种主要的类型：其一是全局（global）还原论，另一个则是局域（local）还原论。全局还原论认为，你自身对世界的经验会逐渐地教你认识到，一般而言，证言是不错的知识来源。比如，你是一个询问火车站方位的年轻人，有人来为你指点具体的位置，尽管你并不知道别人说的对不对，你却可以按照指示自己去验证证言的真实性。既然你总是可以对别人所说的话做再检验，那么一段时间以后你就会从过往证言的记录中找到知识，而这又进一步成为基于经验的理由，促使你去接受现在遇到的证言。对任何一个普通的成年人而言，只要别人告诉他火车站就在这条路一直走下去再右拐，那么他就能马上知道这一点。这并不是因为证言有某种独特的提供知识的能力，而是因为他自己过往的知觉、记忆和推理支持他去接受现在听到的东西。全局还原论者并不需要主张绝对相信任何证言：如果你处在一个存在大量颠覆性因素的境况中，你就需要考虑这些因素——例如，如果你知道自己正在跟一个很有说谎动机的人交流，那么你就要小心了。但是，当没有什么特别的危险信号存在时，全局还原论者就会认为，人们会有站得住脚的积极理由相信自己得到的证言。

局域还原论者则试图把问题谈得更精致化一些：他们没有去寻找某些能够支持信任所有证言的默认理由，而是主张每个人都应该在具体情境中找到特定的积极理由，再去接受自己在所交谈的主题下听得的证言。譬如我们会考虑这样一些具体的理由：证言的提供者是不是专家？他以前告诉过你的是不是真话？他现在给出的说法又有多少可信度？我们所依赖的这些特定理由最终也完全来自知觉、推理与记忆，而非证言自身。如果这些普通理由在给定的情境中足够有力，那么你就能从为真的证言中获得知识。

两种还原论都认为，大多数成年人都能知道自己出生在哪里，也知道世界上存在着南极洲。证言如果是来自亲近和受信任的目击者，譬如你的父母告诉你出生在哪里，或者是来自合适的专家，譬如众多地图制作者与旅行作家告诉你关于遥远大陆存在的事实，那么按照这两种还原论观点，证言都能提供知识。而如果没有信任消息源的具体理由，譬如你在陌生的城市里迷路了，此时你向完全陌生的人寻求方向的指引，那么全局还原论者仍然会认为你获得了知识，而典型的局域还原论者会否认这一点。局域还原论者会说，假如你足够幸运的话，你会得到一个真信念，但仍不是知识。

局域还原论的观点听起来充满了算计的意味，实际上我们并不总是能够在接受证言之前权衡信任与否的理由。然而，关于证言知识的局域还原论并不是一种描述我们事实上是如何在日常实践中形成信念的理论，而是要从理论上刻画使那些信念有资格成为知识的条件。即便我们在问路的时候倾向于盲目地信任陌生人，在局域还原论者看来，我们也不应该由此认为自己获得了知识。如果局域还原论是对的，那么知识的获得要求更多的谨慎和小心。采取这个立场就需要解释，何以基于证言的知识就要比基于知觉和推理的知识需要更多的谨慎小心。的

78

确，证言可能让我们失望，譬如那些给我们消息的人可能不诚实，或者他们自己就没搞明白；但是，知觉同样可能让人失望，有时我们的眼睛会欺骗我们。那么为什么证言就更特殊一点呢？一个可能的理由是，证言会涉及某些有着自己目的的自由行动者。例如，人类的交往与蜜蜂之间的信息传递不同，后者可以以可靠的信号告知彼此哪里能找到有花蜜的植物。如果一只蜜蜂是从其他蜜蜂那里得知花蜜的方位的，那么它就能直接飞到那里去，就好像它曾亲眼看到了那里一样。在这个意义上，蜜蜂的信号表达把彼此经验的好处分享给了其他蜜蜂，有时这个过程被称作“代理性认知”（cognition by proxy）。即便如此，蜜蜂的信号也可能有缺陷，譬如这只蜜蜂生病了，或是有花蜜的植物随后被研究员移动了位置，但这些缺陷就像是我们的感官缺陷一样——当我们生病或是物件在我们不知情的情况下被人移动时，我们的认知也会出错。然而，蜜蜂可以从其他蜜蜂那里直接地获取经验，是因为它们不会故意地欺骗彼此。因此，局域还原论者可能会强调，对于如我们这般狡猾的人类，彼此之间的交往与信息传递就需要格外小心谨慎。

或者，局域还原论者也反对说我们常盲目地听从陌生人的建议。也许我们通常确实都是谨慎小心的，只不过往往感受不到，因为我们并没有明显地权衡信任别人的那些理由。最近有一些对“知识性戒慎”（epistemic vigilance）的经验研究，极大地推进了我们对实际上如何接受他人证言的理解。即便我们没有明确地思考那个被我们问路的陌生人的话是否可靠，我们也还是会密切注意他的面部表情和言谈方式，以评估他是否值得信任。那么，局域还原论究竟是不是一种十分合乎真实情况的理

79

论呢？抑或是它指出了真实情况的草率，我们根本就没有那些我们自以为通常拥有的知识？这就必须要靠对真实情况的更好洞见来帮助我们做判断。

证言：知识的一种独特来源

还可能有一种更为一般化的证言理论。这种理论没有把证言当作依赖于其他认识路径的东西，譬如依赖于过往的经验与推理，而是认为证言本身就是知识的基本来源之一。这就是证言的直接观点，有时也被称为“默认观点”。按照这种观点，如果一个消息灵通的同事告诉你史密斯得到了那份工作，那么“史密斯得到了那份工作”就成为你的知识，而且你的这项知识并不依赖于你的任何推理——比如关于该同事的可靠性及他以往的证言记录的推理。为了获得这项知识，你所需要做的仅仅是理解这位消息灵通人士的话。当然，证言的通道还是要从感觉经验中摄取信息，譬如说你总得能听到别人说的话，能读懂别人写的字；推理也会从感觉经验中提取信息，例如当你看一个完成了一半的数独游戏并且开始作答时，就是这样的。但我们不会因此否认证言是独特的认识路径，正如我们也会承认推理是区别于纯粹知觉的认识路径一样。在理解别人的话时你的思维方式，迥异于你亲自看到某些事情时的思维方式，更区别于你在推理或解谜时的思维方式。

证言的直接观点有着久远的历史。公元2世纪的印度哲学家阿克沙巴德·乔达摩就曾主张过这种观点。乔达摩认为，证言是人们获取知识的特殊路径，强调它并不是任何一种类型的推理。也就是说，我们不会做这样的内心独白：“李先生说史密

80

斯得到了那份工作，而李又是个可靠的人，所以史密斯得到了工作。”恰恰相反，一旦我们理解了李先生说的话，我们也就知道史密斯得到了那份工作。我们也可以关注到这样一个事实：是李先生告诉我们这个事情的，但这并不是我们所获得的主要内容。与局域还原论不同的是，直接观点可以毫无问题地从陌生人那里获取知识：按照古代印度哲学的思路，知识不仅可以从大师或者“贵人”那里直接获得，而且也可以直接得自“野蛮人”，只要他们的确有知识且愿意分享。

当代的直接观点强调，对证言的信任在语言获取和人们的日常交往实践中发挥了重要作用。还原论者和洛克主义者都认为，在人们能够以自己的知觉或推理能力证实证言以前，最好对公共证言采取一种中立的立场；而直接观点的支持者却认为，就这种类型的证实任务来说，我们并没有充足的能力。例如，知道语词的意义非常重要，但只有当别人直接地告知我们某些事情之后，我们才可能获得如此重要的知识。如果我们甚至从一开始就不信任别人会说真话，不接受别人的话的字面意思，我们也就完全不可能相互理解。从这个意义上说，我们的确在某种程度上像蜜蜂那样从别人的话中吸吮着养分。

即便对于那些最慷慨的直接观点来说，证言知识仍有某些必须满足的条件：为了获取知识，你的信息源所说的话必须实际上为真，并且你的信息源也必须知道它为真，而不能只是撞大运的猜测，至少对大多数非还原论者来说都是这样。柏拉图关于这一点有一个生动的例子。他描述了一个能说会道的律师，他试图为他的当事人洗脱一项袭击案的指控。这个当事人的确无罪，但没有目击证人站出来支持他。所以，尽管这个律师自己也

81

不知道他所说的话是否为真，他的确做了很大的努力，以他的个人魅力使陪审团相信他的当事人是无罪的。然后，柏拉图追问道，既然陪审团成员完全是受这位魅力型律师的影响，那么他们真的**知道**被告是无罪的吗？答案似乎是否定的。你要想从某个信息源那里获取知识，那个信息源他自己就得先知道事实。

詹妮弗·拉基是当代主要的证言理论家之一。她把证言知识要求"从已知者获取知识"的条件形象地刻画为"传递水桶的一字长蛇阵"（bucket brigade）。"为了给你满满的一桶水，我必须先有一桶水才能传递给你。此外，如果我给你的是满满一桶水，那么在不考虑洒出水的情况下，在完成这种传递后你所得到的也应该是满满一桶水。"这里，"洒出"的情况就对应于你没听到我说的话，或是有人恶意中伤我是说谎成性的骗子，而你就因此对我起了疑心。不管是古代的还是当代的直接观点，它们都认为，只要你对证言的真理性有所怀疑，那么，即便提供证言的人再有知识，即便你的怀疑事实上根本不合理，你就已经丧失了由证言获取知识的畅通路径。

我们只能从已经有知识的人那里才能获得证言知识吗？拉基对此表示怀疑。设想一下，一位教师对于她被要求讲授的科学理论有所怀疑——譬如说，这个老师是相信"地球还很年轻"的创世论者，而她所讲授的却是自然选择的进化论，照此理论，人类是进化而来的产物，与其他灵长类动物有共同祖先——那么，这个老师就并不真的拥有"自然选择理论为真"的知识，她甚至根本不相信这一点。但是，她却按照国家课纲的要求把科学理论不遗余力地传授给课上的学生。如果信任这位教师的学生在其证言的力量下相信了进化论，难道我们不能说这些学生 82

获得了关于自然选择的知识吗？如果拉基是对的，那么在这样一个例子中，一个自己都还没有“满满一桶水”的人传递出了比自己所拥有的更多的知识。

当然我们还能找到其他类似的方式，都能让人们传递出比自己所知道的更多的知识。一种办法是与他人合作。现在，我可以说自己知道“威拉米特河在美国西北部的俄勒冈海岸山脉和喀斯喀特山脉之间向北流去”。之所以知道这一点，是因为我从维基百科上查到了这个知识，而且做了反复确认。也许正是某个真正知道这一事实的人首次更新了维基百科上的页面，那么也就是真的有充分知识的人通过互联网把他所拥有的知识传递给了我。但也有可能是，那个更新维基页面的人自己也不完全知道这一点，譬如说他可能对岸边山脉的名称并不很确定。然而，一段时间以后，这个词条经过很多人的审核补充，而这些词条编辑人员作为一个群体保证了山脉名称的最终报告的正确性。所以，这个群体也许成功装了“满满一桶水”，也即拥有了充分的知识，共同保证了维基的词条能够给读者提供知识。如果信息来源的可靠性是真正重要的东西，那么在恰当条件下的通力合作要远比单个作者表现得好。

这个例子可以按照不同的理论做不同的处理。例如，洛克主义者会说，我们从维基百科中得到的有关威拉米特河的内容，其实只不过是很可能为真的意见，并不是知识，即便有再多信息源告诉我们完全合理的事实，情况也依然如此。而一个还原论者会说，我们从这里获得的任何知识都是推论性的：按照这个论调，只有对那些已经意识到维基百科的词条是可靠的人来说，相关的词条才真的传递了知识。例如，如果你读过2005年《自然》

83

图7　过去几代人的痛苦

杂志上的一篇文章，其中报告说维基上的词条就准确性而言堪比《不列颠百科全书》，又或者你曾经亲自验证过维基词条的准确性的话，那么那个关于威拉米特河流向的词条就的确传递给了你知识。如果是直接观点的支持者，他会认为，即便对那些尚未认识到维基百科准确性的人而言，只要它的质量控制体系运转正常，它就可以提供关于事实的知识。按照这个观点，任何一个准备学校报告的12岁少年都可以由此得知河流两岸的山脉名称，只要阅读威拉米特河的维基词条就足够了。就直接观点本身而言，如果是传统立场的话，它会要求那些撰写或编辑这个词条中关键词句的人知道所报告的事实；然而，现在的直接观点也可以做出某些调整以便容纳这样的事实，即我们不是从单个的已知者那里获得知识，而是受惠于某个匿名的群体，其中每一个个体都只有部分的而非全部的知识。随着证言的知识论结构研究日益进展，还会涌现更多的信息渠道，以提供新的机会让相互对立的理论各自解释知识的社会传播，从而相互竞争下去。

一般而言，知识论都是在考察完了知觉和推理以后才会转向讨论证言，但也有一些哲学家把它放在通向知识的核心位置上。在这方面最值得注意的是英国哲学家爱德华·克雷格。他认为人们之所以需要知识的概念，目的就是为了处理证言问题：我们把那些能作为好的信息源的人标记为有知识的人。克雷格的这个观点来自这样一个前提：所有生物都企图努力生存下去，而这就需要有与其环境相关的真信念。如果我们不受限于自己所经历过的东西，而是也能够从别人那里学习的话，那么这将极大地帮助我们生存下去。所以我们就迫切地需要有一种能够辨别好的信息源的方法，以便能让这些好的信息源成为我们的耳目；同时我们也需要鉴别出那些坏的信息源，以便知道哪些人的话有可能把我们引入歧途。由此，好的信息源就被确认为有知识的人。

这样看来，克雷格掉转了通常的解释方向：大多数知识论专家认为，成为有知识的人是成为一个（潜在的）好信息源的先决条件；与此相反，克雷格认为，“好信息源”乃是一个更为基础的观念，正是它解释了知识概念的价值及其起源。克雷格的批评者强调说，有知识的人有时也可以是坏的信息源，因为他可能要保密或是故意欺骗别人。此外，克雷格的理论还难以解释另外一些关于知识的直觉。例如，为什么盖梯尔反例中的认知者就没有知识？这用克雷格的理论就很难解释。因为在盖梯尔反例中，那些认知者也可以在某种意义上是好的信息源——作为一个拥有得到辩护的真信念的人，认知者的确可以提供正确的信息，并且在某种意义上他的思考也是可靠的。同时，克雷格理论的进化论维度也是可以存疑的。譬如说，对与我们亲缘关系最近的动物近亲的研究表明，它们也可以区分有知识与无知的状态，但不是在克雷格所认为的更为基本的意义上做的区分。在实验中，黑猩猩无从辨别哪些是能告诉它们食物位置的有知识的信息源，哪些又是对此一无所知的信息源。但是当存在彼此之间的资源竞争时，黑猩猩的确能够追踪谁有哪些知识。例如，低等级的黑猩猩就非常善于记住地位更高的猩猩是否知道食物的位置。由此看来，知识与行为之间的关联要比知识与“好信息源”之间的关联更容易确认。尽管证言的确是知识论的重要主题之一，它却未必会作为我们所有讨论的出发点。

85

86

第七章

标准转换？

考虑语境

有些词的意思是游移不定的。例如“明天”这个词，在每天晚上都会向前移动，指向这一周的另一天。“这里”指的具体地方要取决于你正站在哪里，“我”指的是谁取决于是谁在说话，而“这一个”可以指任何东西。至于像“大”和“小”的概念就更微妙了：一只病态肥胖的老鼠在某种意义上是大的，但在另一个意义上又仍然是小的。那么当我们说“知道”这个动词的时候，是不是也有类似的情形呢？它的意义是否也可能以某种独特的方式游移不定呢？

上一段中所提到的所有这些语词，都有一个共同的特点，就是语境敏感性。对于这些词来说，语境在决定其意义的过程中扮演了重要的角色。像“我”和“现在”这样的词，它们的意义对说话者的身份与所处的时间空间就非常敏感。而像“大”与“高”这样的词，它们的意义对与之相对照的种类非常敏感：譬如我们说一栋摩天大楼很高，跟说一盒麦片很高相比，这个实际需要的高度是完全不一样的。更复杂的是，同一个东西可以是两个不同种类的个体：比如它既是老鼠又是动物，作为老鼠来说它是大的，但同时它又是小动物。那么我们究竟应该用什么词来形容它呢？这完全取决于我们在心里是拿什么种类与之相比较。87

也许有人会说，语境敏感的语词的词义总是在变化之中，但这种说法也不完全准确。为了理解“明天”究竟是什么意思，我们并不需要每天换一本新的词典。因为这里有某些确定的规则。例如，“我”就总是指那个正在说话的人，而“高”总是意味着“在它所属的种类里垂直方向上较大”。所以，语境敏感的语词并不是改变自己的意义，而是像有一个固定的配方那样，总是从交谈的语境中获取某些信息，以便确定它们实际指称的东西。一旦语境确定了下来，像“这一个”或“昨天”这样的语词指称也就随之而明确了。

“语境主义”（contextualism）就是指这样一种观点，它主张像“知道”和“认识到”这样的语词也是语境敏感的。之所以有人愿意接受语境主义，原因之一就是它旨在协调本书第一章和第二章中谈到的一些主要立场。第一章表明“知道”是人们最常用到的动词之一，当我们看到、听到或记起某事如何的时候，通常都会给一个这样的标签说我们知道这件事。当然，你知道你现在正读着这本书。第二章则表明，我们也很容易怀疑人类是不是真的能知道那么多事情。譬如说，你真的能知道自己现在正在读书而不只是在做梦？若要按语境主义者的话来说，上述两种立场其实完全可以是不冲突的。因为“知道”一词本身就是语境敏感的，所以一方面我们可以在日常意义上说你知道很多事情，另一方面我们也可以在怀疑论的立场上主张你其实什么都不知道。这两方面的主张完全可以相容，因为日常的说话者与怀疑论者处于不同的交谈语境之中，所以当他们说“知道”的时候表达的内容也会不同。对于同一个句子，譬如说“明天星期五”，有时说这句话为真，有时说这句话就为假，这取决于你在何时说这句话。同样的道理，我们在做知识归赋的时候说“某人知道自己在看书”，也可以有两种情况。一是在日常的较低标准下，某人的确是知道这一点的；但如果转换到怀疑论者所要求的高标准下，那么可能就得说“某人其实并不真的知道自己在看书”。

语境主义的兴起

作为一种知识论的理论，语境主义发端于1970年代早期的“相关选项”理论。这一理论的支持者认为，当我们说“知道”或“不知道”什么时，总是意味着做某种对比和比较。有一个例子：简带着儿子去动物园，看到前方的围场里有一只黑白条纹相间的动物，于是她就对儿子说：“比利看那儿，那是斑马！”她的确说对了，简的视觉没有问题，能辨识出普通的动物，而且她所看到的那只黑白条纹相间的动物也的确就是斑马。那么我们是不是能说简知道自己看到的动物是斑马呢？我们很容易会说她知道。但一个更棘手的问题就随之而来了：她怎么能确定自己所看到的动物不是一头巧妙伪装过的驴子呢？（图8）驴子也可以被涂上黑白相间的条纹，装饰一下耳朵和尾巴，使得从简的角度看，这头巧妙伪装过的驴子与真正的斑马几无可辨。按照相关选项理论，简所拥有的证据已经足够帮助她做出这是一匹

图8　一个男人与玛拉动物园的一头被涂装过的著名驴子站在一起。安娜塔西亚·泰勒-林德摄于巴勒斯坦自治区的加沙市，2009年12月，来自第七图片库

斑马的判断，这一判断所暗含的对比意思是说，它不会是一头狮子、羚羊或是骆驼等等。在普通动物园所能有的动物范围内，简有一个相对简单的"相关选项"的集合，便于她从中挑选适合的对象。然而，她的证据却不足以使她辨别出这只动物不是巧妙伪装过的驴子：为了做出这个更难的判断，珍妮必须把"它实际上**是**一头巧妙伪装过的驴子"也纳入自己的相关选项集合中，进而再排除这个选项。这对她来说并非完全不可能，但为了排除

这个诡异的可能性，简可能就得跳进关这只动物的围栏之中，走近一些，在动物的皮毛上抹些清洁剂看能否擦掉那些条纹。所以，在二十多步开外的距离处，简知道那是一匹斑马，但并不知道它不是一头巧妙伪装过的驴子。

在这个例子中，像“巧妙伪装过的驴子”这样的东西，不会是任何正常动物园的“相关选项”，至少对简所参观的这家动物园来说不是。但如果动物园经费紧张，或是就爱搞些恶作剧，那么它就可能是这些动物园的相关选项。所以，摆在相关选项理论面前的一大挑战是，如何为任一给定的判断确定它的相关选项？这里面有没有什么规则？这些选项的可能范围部分由对说话者来说实际可能的东西确定，但还有更多的理论细节有待于发掘。比如，简是不是知道这家动物园爱搞恶作剧，与选项的确定是否有关？这个就还不清楚。但更重要的是，相关选项理论有某些深刻的、**不同寻常**的东西涌现出来。

89

假设简知道一些基本的生物学知识，即她像大多数成年人一样知道斑马与驴子是两种不同的动物，所以不可能有一种东西既是斑马又是驴子。那么，按照这些基本的生物学知识与逻辑知识，如果那只动物是斑马，那它就不会是驴子。而逻辑学也会告诉我们，假如某个东西不是驴子，那它也就不可能是巧妙伪装过的驴子。相关选项理论所主张的是，简站在离斑马有二十多步的地方时，她知道自己看到的动物是斑马，同时也知道，如果它是斑马的话，那它就不会是一头巧妙伪装过的驴子。然而，她却并不真的知道那只动物不是巧妙伪装过的驴子。她可以知道方框6中的论证前提，而无法真正知道其结论，尽管她能看到前提与结论之间的逻辑联系。因此，相关选项理论违背了所谓

90

的“封闭原则”——根据这一原则，从你已知的东西中按照逻辑演绎出的任何东西，都是你所知道的。

方框6 知识的相关选项理论的问题

简在二十多步开外就知道的是：

- 那只动物是斑马。
- 如果那只动物是斑马，那它就不是巧妙伪装过的驴子。

而简并不知道：

- 那只动物不是巧妙伪装过的驴子。

违背封闭原则看起来十分奇怪：你难道不信任逻辑演绎吗？更重要的是，一旦我们说简不知道那只动物不是巧妙伪装过的驴子，那么再坚持认为她仍知道那是斑马，就会听起来十分怪异。然而，的确有某些原初的洞见会迫使我们不得不诉诸相关选项理论：因为一方面，说简知道那只动物是斑马，这听起来是正确的；另一方面，说她在远距离处并不真的知道那只动物不是伪装的驴子，也是对的。

91

语境主义就是在这样的背景下兴起的。它一方面保留了相关选项理论的合理部分，另一方面又试图避免违背封闭原则。盖尔·斯泰恩在1976年发表了第一份对语境主义的清晰表述，包含两个部分。第一，斯泰恩主张，我们在不同的环境中采用高低不同的知识标准：“知识概念的本质特征在于，更严格的标准适用于不同的语境。譬如大街上的偶遇与教室或法庭上的环境，会是完全不同的标准——谁又能说在哲学讨论中就不是另一种标准呢？”对于做出判断的那个人来说，并非有一个确定的相关选项的集合。斯泰恩提出，当我们在评论某人是否有知识时，我们考虑的可能是许多或宽或窄的选项集合。第二，斯泰恩主张，在任何给定的语境中，我们都会恰当地坚持某一套标准。譬如说，假如我们担心伪装的驴子可能出现，那么说“简知道那只动物是斑马”就错了。如果简并没有我们这样的担心，她当然可以说，“我知道那是斑马”，而且也是真的判断。但此时简就不能用逻辑学和生物学知识来演绎出更遥远的结论，说它不是一头巧妙伪装过的驴子。只要简开始思考伪装驴子的可能性，那么她所使用的知识标准就必须把这些遥远的可能性也纳入进来，于是她就不能再说“我知道那是斑马”了。不论是在高标准还是在低标准下，谈论知道与否都是合理的；唯一要注意的是，你不能于同一个语境中在两种标准间来回切换，除非你首先转换谈论知识状态的语境。

在斯泰恩的语境主义表述中，知识总是意味着在或宽或窄的相关选项领域中做区分。在日常交谈中，说话者想知道三明治是鸡肉的还是金枪鱼的，那么只有当他能排除另一个选项，确定其中一个时，我们才能说他是对此有知识的。而对于怀疑论者来说，他想要考虑的相关选项就更为遥远了，譬如说，有可能有某些新品种的金枪鱼看起来很像鸡肉，又或是根本没有什么三明治，只有三明治的三维全息影像，甚至是来自某个恶魔的投射在欺骗我们，等等。一旦我们把相关选项扩展到如此宽泛的领域，那么任何人都很难说有真正的知识了。所以，语境主义最直观生动的表达形式，就是相关选项领域的扩展与收缩。但它背后的基本观念实则更为笼统，也并不一定要与“相关选项”的概念连在一起：这个观念主张“知道”的含义会随着语境的转换而变化，而语境主义的各种理论就是在提出决定这种语义转换的不同路径。内在论的语境主义观点是，不同语境要求证据的多少不同；而外在论的语境主义则主张，信念形成机制必须随着标准的升降而在或窄或宽的条件中追寻真理。语境主义就其本身来说并不是知识论，而是一种关于知识归赋语言的理论，是一种可以应用于各种知识论立场的语义“配菜”。

那么，这会是一种让人胃口大开的配菜吗？语境主义的确承诺它能对怀疑论问题给出简洁的解决方案，也就是让争论的双方在某种意义上都是正确的。大街上随便碰到一个人，他可以说“我知道自己正在读书”，而且也的确为真，但怀疑论者说“大街上的这个人不知道自己拿着一本书”，这也是对的。这里的奥妙就在于“知道”这个词为每一个说话者挑选了一个关系。当我们讨论怀疑论时，我们就已经在使用较高的知识标准了；这正像我们讨论篮球运动员时，说他长得“高”的标准也已经提高了很多。语境主义者很喜欢用这个类比来说明问题。譬如，一个身高六英尺（183厘米）的男子通常在美国算高个子了，因为美国男性的平均身高只有五英尺九英寸（175厘米[①]）。但即便如此，到了美国全国篮球协会（英语缩写NBA），六英尺就算不上高了，因为那里的平均身高就有六英尺七英寸（201厘米）。克里斯·保罗就是六英尺的篮球运动员，曾效力于洛杉矶快船队。篮球迷们在评价此队的强弱时可以诚实地说，“克里斯·保罗并不高”；同时，克里斯在交友网站上介绍自己时，也完全可以诚实地描述说自己是个子高的男人。其中并没有什么真正的冲

① 五英尺九英寸约合180厘米，原文应有误。——编注

93 突，因为标准是完全不同的。

这样，当我们要评论别人所做的判断时，就要特别小心了。如果仅仅根据自己的身高超过了美国男性的平均身高，克里斯就说“那些篮球迷说我的话都是假的”，那他自己就先犯了个错。同样的道理，如果篮球迷看到克里斯的交友信息上描述自己个子高，就说他讲了假话，因为他作为专业篮球运动员来说并不算高，那么这些篮球迷也一样犯错了。如果别人说的话是语境敏感的，那么当你去判断其真假时，就需要尊重这些说话者所处的具体语境。同样地，他们也需要尊重你所处的语境。最直接的办法就是把那些潜在的对比类明述出来。例如，“克里斯·保罗的个子高”这句话不完整，留下了一个空白：他相对于什么来说是个子高？这就要由语境来做填充了。我们可以说，“克里斯·保罗是个子高的美国男人”，也可以说“克里斯·保罗是个子不算高的篮球运动员”，这两句话都已经填充了空白，于是也都在各自的语境中表述了真理。

现在让我们回到知识的问题上，要怎样说才最清楚呢？我们是不是能把标准说清楚？“你知道自己正在读书”，是不是就像“克里斯·保罗个子高”一样？如果是的话，我们就有办法解决怀疑论的问题了。因为我们可以这样说：“你看，按照低标准来说，你的确知道自己是在看书了；但要是按高标准来说，你还真就不知道自己在看书呢。”这当然很简单了，但是否能令人满意呢？

其实也很难讲。因为说“高”这个词是语境敏感的，似乎没什么争议；但要说“知道”也是语境敏感的，恐怕就不那么容易了。语境主义经常面对的一种反驳意见是，它只不过是一种改头换面的怀疑论立场。因为怀疑论者本来就不需要挑战日常生活中的低标准的知识概念；在谈论日常对象时，怀疑论者也像其他人一样每天都在不加疑虑地使用“知道”这个动词。怀疑论者真正想说的是，只要我们更仔细小心地反思知识，那么我们就会明白，日常的知识断言其实都是假的。这并不意味着要废除所有对知识的谈论，因为在我们的日常交流中随意地谈论自己知道什么或不知道什么，或许会传递出有用的信息。例如，听你的同事说“小李知道谁拿到了那份工作”，这仍然是有意义的。但是，考虑到“知道”的严格含义，我们也许会得出结论说，小李其实并不真的知道谁得到了工作——他能保证自己就一定对吗？他能保证那个看似成功了的候选人不会被一块偶然飞来的陨石砸中吗？或许我们平常说“小李知道某事”，它真正表达的意思不过是说小李能告诉你一件很有可能为真的事。某些在字面意义上为假的话也可以用来传达有用的信息，这并没有什么不可能的。比如，你会在不严格的意义上使用很多词，像是说“我快饿死啦”，实际上你只是想说自己饿了而已。尽管从字面意义上说这句话是假的，但它还是能起到作用，因为主人听到你这么说就会奉上茶点。在怀疑论的语境中，语词意义的标准提高了，“小李知道谁拿到了那份工作”，这句话在高标准的“知道”意义上就是假的，即便我们在日常语境中说这句话是有用的。但如果语境主义认为，当怀疑论者否定日常生活中人们对简单事实的知识时，譬如说你根本不能知道自己现在正在读书，怀疑论者的确讲出了某些真理，那么至少看来语境主义站到了怀疑论的一边，而牺牲了常识上的知识概念。

94

实际上，语境主义要比上面描述的理论精巧得多，它并不简单地认为日常的知识概念都是“不严格的使用”。特别地，语境主义者非常在意尊重说话者所处的语境。按照这种观点，一方面怀疑论者可以说，“小李并不真的知道是谁得到了工作”，而且这在他们的语境中也的确为真。另一方面，日常生活中的普通人也可以说“小李知道是谁得到了工作”，怀疑论者在日常语境之中也**不能**说这句话为假。这个道理从另外一边来讲也是一样：普通人可以说“我知道自己正在看书”，这在日常语境中是个真语句；但他们也不能说怀疑论者就是错的，因为在哲学讨论中需要用到较高的知识标准，而在这个前提下，“你并不真的知道你在读书”，这样说也没什么问题。语境主义者并不需要在怀疑论哲学家的语境与常人生活的语境之间做孰优孰劣的区分，因为高的意义标准并不必然优于低标准。实际上，如果你只是想要写一张贺卡祝贺某人得到了工作，那么那个高的知识标准显然就不仅惹人生厌，而且也过于迂腐了。只要这两方各自安好，不贬低另一方的断言意义，则会是一个双赢的局面。

然而，语境主义对其他观点的这种和风细雨般的包容精神并不能打动所有人。批评这一观点的人坚持说，怀疑论与常识的立场不可能都是对的，其中必有一方才**真正**说出了有关知识问题的真理。一旦确定了谁是正确的一方，我们就不再需要其他的立场了。而从语境主义的角度看，想要问究竟是哪一方才真正抓住了知识的本质，就像是问到底星期几才是真正意义上的“明天”一样，本质上是毫无价值的问题。

如果语境主义都以友善的态度面对怀疑论与常识的立场，那么，对于“小李是否真的知道谁得到了工作？”这个问题，相信它不管在什么语境中都有唯一正确答案的人，就是语境主义在哲学上的对手。而语境主义面对它的这些对手时就不能那么友善了。语境主义者会说，这些人实际上犯了个错误，而且也不只是他们，如果有人认为怀疑论与常识立场的话不可能同时为真，那么持这种观点的人就也错了。语境主义者的确认识到，通常人们都会认为必须要选一个立场。毕竟，相对于“高”或“这里”这些词来说，“知道”的意义转换似乎更加让我们难以捉摸。所以，对于到底哪里算“这里”的问题，就不会有什么宏大的哲学传统来处理相关争论；但对于什么才是真正的“知识”，哲学上的争论就有着非常悠久的传统了。既然我们在谈论时间、地点或是像“高”这样的性质时，其语境的敏感性都是非常显而易见的，那么，为什么到了谈论知识的问题时，语言功能上的语境敏感性就对我们隐而不彰了呢？这个问题是当代语境主义者最为活跃的研究主题之一，他们也提出了许多解决的方案。一种可能性是这样的：我们在使用“知道”的时候，实际上关闭了进一步的探究，从而阻止了我们深入追溯这一概念意义的语境转换，而我们本来应该继续这样做的。这样就造成了一种幻象，好像知识本来就是绝对的。另一种可能性是说语境主义就是错的，知识的确是绝对的。说知识是绝对的，意思是说我们使用的相关语词并不是语境敏感的，这种观点被称作“不变论”（invariantism）。而它要面对的挑战是，如何解释使得知识时而容易时而困难的转换的直觉？ 96

利益相关的不变论

小李下班后在去公交车站的路上，偶遇了他的同事史密斯，那时史密斯正要回单位大楼。

——“小李，你知道一楼供应室的门锁了没有？我刚想起来我的外套落那儿了，我也没有钥匙。”

——“是锁了的——就是半个小时前我锁的门，所以我知道，而且之后我也没在过道里遇见别人。抱歉！”

这里，小李主张自己真的有知识，似乎并没有什么争议。我们可以进一步假设，实际上那个门的确是锁了的，而小李的钥匙、视力和记忆也都没有问题。但现在来设想这个故事的另一个版本，在这个版本中，小李离开大楼后发生了另一个转折：在他走向公交车站的路上，小李碰见的不是史密斯，而是四位警官。

——“先生你好，对不起打扰了，但目前有个紧急情况，就发生在你刚刚离开的那栋大楼里。它的二楼发出一声枪响，开枪者目前还在大楼里。除了前门以外，还有什么别的途径能够离开那栋大楼？”

——“供应室那里有一个后门，但我在半小时以前已经把通往供应室的门锁上了。”

——“你真的确定它仍然锁着吗？有没有可能被别人打开过了？”

——“这我就不知道了——我没看到有什么人经过，但我也没一直盯着那扇门。”

在落下外套的故事里，小李声称自己知道门被锁了；而在枪击案的版本里，他却说自己并不知道有没有锁着门。这两次他主张的东西听起来都是真的。奇妙的是，这两个故事竟然可以

97

与小李离开大楼的那个时刻并行不悖：在这两个情境中，小李都信任他在这半个小时内的记忆，并且用回忆的内容来回答自己是否真的知道门被锁了的问题。在传统知识论中，你是否真的拥有知识，取决于一些传统因素，譬如信念的真值、证据的优劣等。有趣的是，所有这些因素在上述两个故事里似乎是完全等同的。那么，在这些前提下，怎么可能小李在前一个故事里知道门被锁了，而在后一个故事里就不知道了？

语境主义者的解释非常明确：在公交车站进行的随意交谈里，实际起作用的是知识的较低标准，而当警官问话时，知识标准就被提高了。所以，小李在第一个故事里说自己知道是没问题的，在第二个故事里说“我不知道”也没有问题。但这个解释实际上并没有告诉我们，小李是如何在第一个故事里拥有知识，而在第二个故事里就没有知识了的：这其实是关于同一个人在两个语境下能说什么真话的解释。语境主义者会坚持说，从更高的知识标准来看，在第一个故事里，说小李真的知道门被锁了，其实是**假**的。所以，对于“小李究竟是知道还是不知道？”这个问题，就没有一个独立于语境的简单答案。

如果我们觉得，小李在第一个故事里真的知道那个门锁了，而在第二个故事里又的确不知道这一点，那么，这种感觉会把我们推向语境主义，以便找到某种能直接合乎这种感觉的知识论立场。如果知识论中的所有传统要素，像是真、证据、可靠性以及其他诸如此类的东西，都在两个故事里完全等同的话，那么剩下的一种可能性就是，有某些非传统的因素造成了其中的差异。那么，在这两个故事里，有哪些要素是不同的呢？一种被称作“利害相关的不变论”（interest-relative invariantism）的观点认为，这两者之间存在着某些**实践**上的差异。在第一个故事里，小李并没有面对多么利害攸关的选择。为了让史密斯能取回他的外套，作为一个好心的同事，小李会走回去帮他开门。但即便他在这个事情上搞错了，也没什么大不了的。走回去才发现门是开着的，那也不过就是浪费了几分钟时间多走了几步路而已。而在第二个故事里，一旦小李在"门是否锁着"这件事情上搞错了，实践上的后果就变得非常严重了，因为那个持枪者就有可能从后门逃之夭夭。

利害相关的不变论者发现，实践上的利害在很多情境中都发挥着影响。譬如说，你知道那个三明治是鸡肉的还是金枪鱼的吗？如果这并不是什么利害攸关的选择，就像你只是稍微更喜欢鸡肉三明治，而如果是金枪鱼的话，你就会倾向于不要三明治而换成汤了，那么一番随意的查看就足以主张自己知道了——譬如说，看到它像是鸡肉的，就可以说知道那是鸡肉三明治。但是，假如这个选择对你来说非常利害攸关，譬如说你对鱼肉有严重的过敏，那么仅凭随意的查看就不足以声称自己有知识了。所以，按照利害相关的不变论观点，你的选择越是利害攸关，你就越是需要更多的证据才能主张自己有知识。这也就是为什么小李故事的两个版本会在"小李是否知道"的问题上得出完全相反的答案。

利害相关的不变论区别于语境主义的地方在于，它是关于知识本身如何运作的理论，而不是关于知识归赋的词汇的语义学理论。它所做的判决并非仅在某些语境的表达中为真，而是彻底地为真。小李在第一个故事里有知识，而在第二个故事里的确没有。在这些判决中，是语境决定了小李拥有知识需要有多少证据的标准，而这最终是小李自己的语境，而非那些可以从不同的视角谈论小李的人所处的语境。按照这种不变论观点，当怀疑论者说，即便在第一个故事里，小李也不知道门锁没锁，那么他就搞错了。不变论的支持者尽管也对知识的本质有不同的见解，但他们有共识的一点是，正是实践上的利害参与决定了人们是否拥有知识，而这个要素不在传统知识论学者的考虑范围之内。随着风险上升，知识所需要的证据就会更多。支持这一理论的学者主张，利害相关的不变论最好地阐释了知识与行动之间的关系。

又是过时了的不变论？

语境主义者很快开始批评这种不变论，因为他们注意到，利害相关的不变论同样难以把握直觉上的某些转换模式。假如真的像不变论者所主张的那样，在第一个落下外套的故事版本中，小李的确知道门是锁着的，而且这也是一个平淡无奇的、独立于任何语境的事实，那么，怀疑论者又何以能如此容易地让我们开始怀疑这个事实呢？当我们谈论反事实条件的可能性时，利害相关的不变论观点就显得更加捉襟见肘。例如下面这句话："侍者并不知道那个三明治究竟是金枪鱼的还是鸡肉的，但是，假若不是因为他的顾客中有人对某个选项过敏的话，侍者本来是知道的。"这听起来很奇怪，但利害相关的不变论者会说，其实这也是可能的。同时，支持不变论观点的人还会反过来批驳语境主义观点，通常他们会指出，"知道"这个词并不像其他语境敏感的词汇一样。例如，"知道"不像"个子高"，它不会很容易地适应不断变化的尺度；"知道"也不像"今天"，没有什么简单的规则能像说清楚“今天”的含义那样，说清楚“知道”如何受语境的影响。语境主义者已经表明，“知道”很可能有某种特殊的语境敏感性，但究竟它是怎样起作用的，仍然是未定之数。

正当主张标准转换的不同观点之间打得不可开交之时，还有一些人支持更为刚性的知识标准，希望知道他们那种过时了的观点在今天是否能赢得一席之地。严格的不变论者主张，知识就是由传统知识论的那些要素所严格决定的，包括真、证据和诸如此类的东西。在这个意义上，用作知识归赋的词汇就不再是语境敏感的。实际上，怀疑论就是严格不变论的众多直接形式之一。例如，学院派怀疑论认为，要拥有知识就必须符合某个唯一确定的标准，这就是要求我们对所判定的对象拥有无可辩驳的正确印象。不幸的是，我们从来没有在日常生活中满足这一标准，也许只有一两个像“我存在”这样的特殊判断能够达到标准。如果你认为自己真的知道你正在读书，那就错了。但是，如果知识离我们竟是如此遥远，而我们对知识又谈论得如此之多，那么怀疑论者似乎就欠缺一个解释，何以会是这样，对此我们可能感到满意或者不满。如果我们感到不满，那也许就会想要转向温和的严格不变论立场——我们可以说，知识的标准的确是唯一的、固定的，但它也是一个人们通常能够达到的标准，就像你的确知道自己正在读书，而小李也真的知道那扇门已经锁了。但这种严格又温和的不变论也还是有它自己的问题：它需要解释为什么怀疑论者能如此轻易地让我们怀疑自己做出的日常判断，而虽然小李在两个版本的故事中所依赖的严格的传统标准是相同的，但为什么他一旦被警官问话，他原来所拥有的知识似乎顷刻间就烟消云散了呢？对这些问题，温和版本的严格不变论付出了很多努力去寻求解答。他们所尝试的方向之一就是论证直觉的转换中出了错，或是造成这种转换的反例本身就有问题。还拿锁门的这个故事来说，其实两个版本之间的差距也许要比它们看起来的大很多。我们原来假设，在这两个版本中，决定知识标准满足与否的传统要素是等同的，但情况也许是这样：高风险将会自然地导致信心降低，或是使当事人以不同的方式思考既有的证据。另一种可能性是，在高风险的情境中，我们模糊了“知道”与“知道我们有知识”的界限，或是没有搞清楚实际上说了什么与想要表达什么之间的差异。我们对这些案例的直觉可能出了某些问题，又或许在与怀疑论对话或讨论生死攸关的话题时自然地发生了某些曲解，等等。考虑到类似问题的困难，可能有必要深入地考察一下，人们对知识的直觉究竟是怎样产生的，这应该是进一步合理解释此类直觉模式的必由之路。

第八章

关于知识的知识

知识论的原初材料

在系统地研究知识问题以前，我们对知识都知道些什么？我们并非白手起家。哲学家们有特别的理由希望，我们从一开始就具备某种识别真正知识案例的能力，即我们对具体案例的本能或直觉，应该会支持关于知识的某一些哲学理论，胜于其他的哲学理论。比如，如果你觉得那个望向停走的时钟的人并不真的知道时间，那么这种感觉就会成为你拒绝知识的经典定义的理由。但是，是什么让你做出了那样的判断？无论别人是否也采取相同的方式，你自己的判断方式究竟是否正确？这一系列的问题最近激发了新的研究工作，其中有的是经验性的，有的是哲学上的，都是为了研究我们关于知识的直觉。

读　心

“直觉”这个词或许表明了某种神秘的洞见力量，但其实日常生活中存在大量的对知识的直觉。我们可以先来想一下下述两个判断之间的区别：（1）“小李认为他被跟踪了。”（2）“小李知道他被跟踪了。”这里面的确是有重要的差异，但其实到底选择用哪一个判断通常并没有经过特别用心的计算。你会模糊地觉得，人们拥有知识并不需要内心先有一个明确的知识论观点。这种感觉就是直觉。读心就是这样一种自然的能力，它使人们产生对知识与其他心理状态的直观感受。“读心”这个词也常用于舞台上的魔术表演，其中魔术师玩了一个看起来根本不可能做到的把戏，就好像真的能读取他人的想法一样。但这个词在心理学家那里完全不是夸张，就是用来指人们在一天中经常会有的经历。所谓读心，就是把某些“隐蔽的”或藏在背后的心理状态归赋给他人，这些心理状态包括需求、害怕、相信、知道、希望以及诸如此类的其他状态。如果你看到某人伸手去够东西，你不会只是看到一只胳膊在空间中伸展，而是看到一个人想去拿盐罐，想要某个东西，以及努力地去得到它。也许只是在下意识中，我们才会注意到人们够东西与匆匆一瞥的方式，感受到其面部表情上的细微变化，从而获得关于其内心状态的线索，理解他们对外部环境的把握，于是我们就能更好地预测他们的行为，以及他们将如何与我们互动。如果没有读心的能力，我们将只能看到表面的肢体动作与面部特征的模式；只有通过读心，我们才能了解一个人内在的状态究竟是怎样的。无论我们是否想要与他人协调一致或相互竞争，知道他人想要什么、知道什么，是否友善、愤怒或缺乏耐心，都将使我们受益良多。当然，我们并不总能得到正确的答案，而很有可能搞错了别人所知道、所想要的东西，或是被某个技巧高超的骗子所蒙蔽。但总体来说，我们

102

日常的社会交往是成功而有效的，我们对情境的误解乃是一种偶发的、会引起惊异的情况，而不会是常态。

人类有着比地球上其他物种更强的读心能力。黑猩猩也会监控对手是否知道隐藏食物的位置，它们能够理解拥有知识与缺乏知识之间的简单区分。但人类还可以追踪别人犯错的
103
路径，而这是目前我们所知的任何其他动物都还做不到的事情。你能够看出别人持有假信念，譬如你在朋友身上搞恶作剧，把他的麦片盒倒空，放进去橡胶做的甲虫。那么你就知道，当他坐在桌前准备吃早餐的时候，他会期望从那个盒子里倒出燕麦片来。你也知道他的这个内在表征并不符合实际的外在事实。任何其他动物似乎都不能表征这种状态，即便是在某些看起来非常需要这样做的情境中。当人们用实验测试动物是否具备追溯其他动物错误信念的能力时，所有非人类的动物一概都未能成功。

这些实验设计得非常难，即便幼年的人类也会失败。例如下面这种不成功的情况就很经典：给被试的小孩拿一个他熟悉的容器，比如上面画着糖果图案的糖果盒子，然后问他这里面有什么。当然小孩会回答说："是糖！"然后意外的事情发生了：打开盒子以后发现里面没有糖果，只有蜡笔。然后再把盒子合上。现在再告诉他，还有另一个一直等在外面的小孩，然后问：当那个小孩走进房间的时候会怎么想？那个小孩会知道盒子里面有什么吗？他会认为里面有什么？几乎所有五岁的小孩都能够回答出，那个在外面的小孩将会对盒子里的东西持有虚假信念，但如果是三岁的小孩，就只有一小部分能够回答正确。很奇怪的是，大多数三岁小孩都会说，那个外面的孩子已经知道盒子里面是蜡笔；更令人诧异的是，如果问三岁的小

孩他们一开始看到盒子时的想法，许多人都会给出错误答案。设想他人拥有虚假信念是一件要求较高的任务，就像回想你自己过往的错误一样困难，因为你需要压制住自己对现实的当下印象，以便能够从那个出错的人的视角去看问题。信念或意见可以符合外部世界，也可以不符合，是一种相对复杂的表征状态。如果知识是一种必须反映事实本身的状态，那么知识的归赋就容易多了。

主张“知识优先”纲领的人试图从这些发现与其他研究中获取对自己观点的支持，那些研究的观点是，知识的状态要比纯粹信念的状态更容易表征。例如，“知识”一词在各种文化中都要比动词“相信”更早习得，也用得更多。批评者们则认为，习得概念的顺序并不表明其中哪一个更基本。毕竟，我们用食用盐这样的概念，要远远早于了解构成它们的成分，而且我们也更经常地提起这一化合物。另一个经常被提出来的疑问是，如果孩童因为年纪小而无法理解虚假信念的概念，那么在同样的年纪是否就能把握知识的概念呢？更麻烦的是，新近的一些研究表明，很小的孩童甚至是婴儿就已经能部分地鉴别虚假信念了，只不过这种能力的范围与意义还不甚清晰。目前，我们仍然在试图搞清楚人类识别知识的能力究竟是怎样发展起来的，而对这一困难问题的持续研究仍会为我们知识与信念的日常概念的自然结构给出日益清晰的图景。

更一般地说，经验研究将会更好地揭示人类读心能力的本质局限。这些能力包含许多高度专门化的装备。方框7中的三个小故事展现了这些装备，它们被用于麻省理工学院的神经科学家丽贝卡·萨克斯领导的研究之中。

方框7　激活不同脑区的小故事

1. 物理

在核桃林边缘的红色大谷仓后面，是附近最壮丽的池塘。它又宽又深，岸边有一棵老橡树。池塘里有各种各样的东西：鱼、旧鞋子、丢失的玩具和三轮车，以及许多其他的惊喜。

2. 人

老麦克菲格尔比先生是一个老农民，有着灰白的头发和皱巴巴的面庞，穿着灰白色的皱巴巴的旧工作服和旧靴子。他一辈子生活在这片土地上，年纪甚至比这里的许多树都大。小乔治是麦克菲格尔比先生在镇上的侄子。

3. 心理

麦克菲格尔比先生不想让任何小男孩在池塘里钓鱼。但小乔治装作没注意。他非常喜欢钓鱼，而且他知道自己跑得比镇上的任何人都快。乔治决定，如果麦克菲格尔比先生发现他在钓鱼，他就会快点跑掉。

在你读这几个故事的时候，你的大脑上的不同区域会有选择性地被激活。对于小孩子的大脑来说，第二个和第三个故事激活的脑区并不会有什么特别的差异，但对大一些的孩子和成年人来说，第三个故事是非常特别的。理解这个故事就需要能表征其中人物的心理状态。像这样的故事就会选择性地激活靠近右耳的脑区——右侧颞顶联合区（RTPJ）。其他的脑区也会参与，像是内侧前额叶皮质（MPFC）就是在成人和孩子阅读所有与人相关的故事时会被激活的脑区，当然也包括了方框中的第二个故事。对小孩子来说，阅读任何有社会性内容的故事都 105 会激活右侧颞顶联合区，但对成年人来说，这个区域只会专门响应与人们的知识、需要、决定、假装和信念等诸如此类的东西相关的故事。如果这个脑区遭遇外力重击或内部病变的损伤，抑或是在实验中因经颅磁刺激而暂时关闭，那么，对于他人知道什么以及会做什么，我们做出评价与预测的能力都会被极大地削弱。

成年人的大脑里有专门负责读心的区域，这并不稀奇。尽管我们可以不假思索地做这件事，但实际上读心所用到的计算 106 是非常复杂的。在这方面它很像面部识别的能力，都涉及非常迅速而不假思索的计算过程，也都在成年人大脑中有着高度专门化的区域。设想一下，当你观察某人的时候，他想要、注意、假装与计划什么之间究竟有着怎样的关联？在这些模式中，操作乃是一项很有意义的工作。在日常的交谈中，如果我们能够决定把某人描述为仅仅相信某件事情，还是说他真的知道了这件事情，那么这都可以在不假思索的计算中完成。之所以能做到这些，某种程度上就是因为，我们的大脑有专门负责追溯他人心理状态的功能区域。

我们的读心能力也有一些自然上的限制。首先，我们所能表征的心理状态的嵌套层级是有限的。下面这句话里用了四个层级："戴维斯认为，小李知道史密斯并不想让琼斯得到关于工作的信息。"你能理解这句话吗？研究表明，人们有可能追溯到九个层级，但是大多数成年人只能做到五个层级的水平，超过这个层级之后，读心能力就会停止工作，人们开始随机地回答理解与否的问题。对于那些社交面广的人来说，这方面往往会做得好很多。这种限制就有点像是追踪屏幕上活动对象的有限能力。大多数人只能同时追踪五个活动对象,资深的电子游戏玩家可能会水平更高:这两类能力的自然限制都在于,我们只有这么多的注意力可资调用。

在读心的能力上,还有更深的一层限制是这种能力独有的,那就是一种自我中心化的倾向。更准确一点说,我们受制于一种"自我中心主义"的偏见。因此,当我们评价那些对自己的境况了解比我们少的人时,我们会难以克服自己固有的视角。在低龄段的孩子中,这样的偏见是非常常见的,前面提到的那个糖果盒子的任务就是其中的一个例子。虽然第一个小孩知道糖果盒里是蜡笔了,但他却难以想象别的小孩子会不知道这一点。儿童的认知通常都会遇到这样的困难,就是难以认识到他人从世界中获取信息的局限性。例如,让一个孩子举起两个存钱罐,两个虽然看起来完全一样,但一个轻一个重。即便这是个大一点的孩子,能够通过糖果盒任务的测试,他还是会错误地预判说,别人隔着老远就能看出这两个存钱罐哪个重。

甚至连成年人也还是会在类似的问题上出错。在推算一个所知较少的人的视角时,我们很难悬搁掉自己对情境的私有知识。譬如说,在炒股游戏中,假如其中有一方得到了"内幕"消息,且他们也知道对手缺乏这些消息。按道理来说,信息上占优势的这一方应该把自己得到的信息搁置一旁,以便能更好地预测对手会对自己的报价做何回应,此时对手的视角由于缺乏信息而更简单,但实际上信息占优势的一方却很难做到完全搁置自己的信息。当我们评价他人决策过程的优劣时,我们也很难避免自己对情境的私有知识的影响:假如你得到的消息说,某一个本来机会渺茫的赌局最终却得到了令人满意的结果，那么这个信息就会影响你评价赌局玩家的决策明智与否。尽管你也认识到，把结果的信息纳入考量并不合情理，但这个影响的程度也还是会超出你的预期。

在表征或评价他人视角的时候，为什么我们如此难以去除自己私有的特殊知识的影响？目前还不完全清楚这其中的原因。但我们可以确定的是，这个自我中心的偏见特别顽固：如果提前警告有这样的偏见存在，或是给以利益刺激，那么有可能会暂时压制这些偏见以获得更好的行为表现，但即便是在这些条件下也还是不能完全去除自我中心主义的影响。知识论学者想要讨论的是，既然这是我们看待他人视角时的自然局限性，那么它是否也在某种意义上激发了那种支持语境主义的直觉模式？一旦我想到了“巧妙伪装过的驴子”这样的可能性，那么我就难以评价一个动物园参观者的简单视角：即便我明确地意识到，这样一个简单的参观者并不会设想那些奇奇怪怪的可能性，但是，自我中心主义也还是会让我以类似的方式去评价他。无论这是否能解释语境主义背后的知识直觉，能够更好地理解这样一些自然机制——正是它们决定了我们对有无知识的直觉判断——知识论学者都会收益良多。如果其中的某些直觉是从读心能力的自然局限或偏见中衍生出来的，那么在构建知识的理论时我们就必须特别小心地加以处理。

案例法面临的挑战

对直觉的本质感兴趣的哲学家可以去关心读心的经验研究，但也可以尝试做某些更为直接的工作：面向普通人对知识论重要案例的直觉做民意测验。这种方法就是所谓的“实验哲学”（experimental philosophy），在它发展的早期确实提供了一些出人意料的成果。实验哲学方面的第一篇论文是在2001年发表的，是由乔纳森·温伯格和他的同事们完成的，旨在研究对一系列小故事的直觉回应，这些小故事都与人们的判断有关。有的就很简单：“假如有人对抛出的硬币将正面朝上有‘特殊的感觉’，那么他是否真的知道硬币会正面朝上？”大概九成以上的回答是否定的。有的案例就很复杂，包含一些复杂的假设性秘密约定。下面的方框8就描述了一个较难的案例，这其实是一个盖梯尔反例。在实验中，绝大多数被试者都报告说，故事中的核心人物并没有知识，判断的理由与主流哲学家的主张相同。但也并非所有亚群体都支持相同的观点：在66个自我认同为“西方人”的被试者中，有74%的人给出了上述标准答案；而在23个自我认同为“东亚人”的被试者中，只有43%的人给出标准答

方框8　温伯格及其合作者的美国汽车案例（2001）

吉尔是鲍勃的朋友，她多年以来一直开一辆别克牌汽车。所以，鲍勃就认为吉尔开的是一辆美国产的汽车。然而，他所不知道的是，吉尔的别克牌汽车最近被偷了，也不知道吉尔已经换了一辆庞蒂克牌汽车，这是另一种美国产的汽车。那么，鲍勃现在真的知道吉尔开的是一辆美国产的汽车吗？还是说他仅仅是相信这一点？

请选择：

真的知道　仅仅相信

案；到了另外23名来自“印度次大陆”的被试者中，这个比例则是39%。由此，温伯格和他的合作者得出结论说，在知识论上信任自己直觉的哲学家们并不是在揭示知识的客观本质，而只是在研究表现他们各自固有文化态度的完全不同的行为。这些其他文化共同体或许并不具有的态度，也就未必会向我们展现知识的本质。所以，实验哲学运动倡导人们不要使用直觉来研究哲学。

这让传统的哲学家们很恼火，他们提出了各种各样的批评。有的批评针对的是这个案例的细节。比如，这个案例的主题内容是美国汽车品牌，不同的文化群体对这一主题的兴趣也有差异，有的一下子就能辨识出别克是美国的汽车品牌，那么对于这些人来说，这个故事就更有可读性，更容易理解。更进一步说，这个故事还可以以不同的方式更加充实和完整起来。例如，我们是否要假设吉尔总是开美国产的汽车，而且鲍勃作为她的朋友也知道这个事实呢？如果从这个角度来重述故事，那么它甚至连盖梯尔反例都算不上：因为，别克牌汽车的被盗并不会改变吉尔这个“只开美国产汽车”的长期倾向，而她的朋友也是知道这个倾向的。还有的哲学家主张，我们不应该过于看重非专业人士给出的不假思索的意见，因为这些意见无疑地会受到各种随机因素的影响；而对哲学图景的直觉之所以重要，乃是因为它是认真反思之后的结果。然而，在认真反思的过程中，谨慎与细心是绝对必要的，至少对那些声称自己在发展理论以理解前理论直觉的哲学家来说如此。但这同样也会招来批评者的指责：如果你还需要对自己的直觉做非常艰难的思考，那么你的判断将难以避免受到所持有的理论观点的污染。 110

温伯格的这项原创性发现最近受到了一些挑战，但同时遗留了一些实验哲学的更大问题以待解决。通过对更广泛的案例的系统性研究，结果表明，哲学家们传统上所说的那些直觉，实际上并不专属于西方人。规模更大的研究从各种不同的文化中选择被试者，在各种盖梯尔反例，包括那个最初的“美国产汽车”案例上，结果发现存在着非常一致的知识归赋模式，不分文化差异或性别差异。普通人的确倾向于赞同专业哲学家们的意见，主张盖梯尔式反例中的认知者并没有真正的知识。温伯格及其合作者曾关心的是，盖梯尔式的回应只是从非常特殊的文化群体中生发出来的，因此并不能作为知识本身之有无的证据；而这些后续的研究更倾向于支持说，这样的回应乃是有着普遍的意义。这些新近的发现符合对知识概念的演进所做的跨文化研究。不论是来自伦敦和东京大型都市幼儿园的儿童，还是来自喀麦隆农村从事狩猎-采集的社群的儿童，他们都经历了同一个把知识与无知或假信念区分开的漫长过程。可以肯定的是，成年人的行为表现在不同文化中有时会有些差异。例如，与美国成年人相比，中国的成年人似乎能更快更准确地推算他人的视角。但在知识归赋的根本能力上，不同的文化共同体之间似乎并没有什么差别。如果说，实验哲学的早期研究加深了这样的焦虑，即哲学家与非专业人士之间并不共享关于知识的直觉，那么，新近的研究则表明，这样的焦虑并无必要。然而，更深层的焦虑依然存在，因为某个印象被广泛地接受并不意味着这样的印象就是正确的。111

正如在盖梯尔反例上普通人与专业哲学家共有一些直觉，在某些更有争议的案例上他们也会有相同的直觉。设想一间普通的家具店里有个人叫阿尔伯特，他看到一张亮红色的桌子非常喜欢，查看了它的价格后，问他太太：“你喜欢这张红桌子吗？”就这个案例来说，你是否能说阿尔伯特知道这张桌子是红的？在新近的研究中，有92%的学生给出了肯定的答案。现在考虑一下这个故事的另一个版本，只需要加上少许的细节。这张桌子的确是红的，而且家具店里的灯光也是正常的，但你要注意到这个更进一步的事实：假如这是一张白桌子放在亮红色的聚光灯下，那么在阿尔伯特看来它**同样**会是亮红色的桌子，而他**并未检查过灯光是否正常**。那阿尔伯特还会知道桌子是红的吗？结果只有不到一半的被试者回答仍是肯定的。由于某种原因，只要想到某种出错的可能性，即便很清楚阿尔伯特根本没有这种出错的风险，仍然会极大地弱化人们对他的知识归赋。需要注意的是，上述两个故事并不是对不同的客观环境中的两个不同人物的描述，第二个故事只是在第一个故事的基础上添加了某些细节，后者本来也可以讲述。这就像是当我们更加认真地思考案例时，我们对待知识标准的态度，从较为慷慨豁达转换为更为苛刻和吝啬。正是这样的直觉模式促使部分哲学家走向怀疑论，而另一部分则求助于语境主义。

表面上看，这两个判断之间有某种不连贯性。怀疑论者认为，我们只能信任第二个直觉，即主张阿尔伯特没有知识；毕竟这是在我们认真思考情境之后拥有的直觉。语境主义者则认为，这种表面的不连贯性并非真实的存在：这两种直觉都没什么不对的，甚至也根本不冲突，因为在第二个情形中，当我们思考看似不可能的可能性时，我们会以更严苛的标准使用“知道”这个词。阿尔伯特满足了较低的标准，但并不满足那个高的知识标准。这正像说一个人相对于普通美国人来说是高个子，但相对于专业运动员来说身高一点也不突出。同时，温和的不变论

112

者会说，第二个更具怀疑论性质的直觉才是不可取的，因为它有可能产生于某些心理上的偏见。例如，我们还可以这样来评价阿尔伯特：他自己实际上已经在思考有异常灯光影响的可能性了，但在判断桌子为红色以前，顽固地拒绝检查灯光是否正常。按照温和不变论的观点，我们对这一案例的回应乃是一种认知上的错觉：大脑中的某些东西在这里自然地估算错了，正像在赫尔曼栅格错觉（图9）中，我们的眼睛会自然但错误地发现，在栅格的框角处有一些暗点。实验哲学的工作表明，这样的直觉模

图9　赫尔曼栅格错觉

式广泛存在，但它不能告诉我们哪一个直觉才真正抓住了知识 113
的本质，也不能说明哪些直觉是错觉。我们怎么才能解决这个问题呢？

知识的印象与知识本身

有的哲学家主张，我们所面对的乃是一个毫无希望的处境。如果怀疑你戴的手表走快或是走慢了，你还可以用美国国家研究委员会的原子钟来检验；但如果你想检验自己关于知识的直觉是否准确，就没有任何明显的对应途径。如果对这些具体的案例产生的直觉回应缺乏一个融贯的知识论理论解释，你可能就会怀疑，其中有一些直觉只是错觉而已。但到底哪些才是错觉呢？哲学家罗伯特·卡明斯主张，假如对于知识自身的本质，能有某些独立的、无须直觉介入的途径去把握，那么我们才有可能分辨出哪些对知识的直觉是正确的。但是，如果我们有这种直接途径的话，我们就不必用对具体案例的直觉来把握知识的本质："如果你的知识已经多到足够让你用哲学直觉来解决问题的话，你也就不再需要直觉的介入了。"因此，卡明斯得出结论说，哲学家不应该依赖对案例的直觉来做知识论的研究。

应该说，这是个非常悲观主义的回应。如果拿知觉来做类比的话，这相当于说：我们必须要先有某种独立的、不依赖于视觉的途径来理解颜色，才有可能解决对颜色的视觉印象的种种问题，才可能分辨出其中哪些是错觉。事实上，我们的确发展出了某些分辨颜色信号的技术手段，以便摆脱对肉眼的依赖。例如，光度计可以测量赫尔曼栅格图上的颜色，而且不会受框角暗点的干扰。尽管如此，我们并不真的会等到有了这种仪器才开

图10 我们能否取得进步?

始区分错觉与正确的印象。长期以来，我们一直用行之有效的手段来分辨哪些印象是正确的，譬如说我们会对获得的印象在不同的环境中、从不同的角度做一再的确认。我们对视觉的理解正是伴随着我们对光与颜色本质的理解一同演进。

同时，即便在知识论上并没有光度计那样的装备可以探究知识的状态，对知识直觉的探究也一样会随着对知识本身的探究一同演进。如果我们能更好地理解分辨知识的直觉能力，使之纳入心理学的广阔图景，我们也就能更好地判断出哪些直觉才是真正重要和有意义的。同时，如果我们能对知识本身提供更为敏锐的哲学构想，就将有助于推进我们对这些直觉本质的理解。直觉以外的工具也可以用来处理知识的问题。我们可以努力发展内在一致的知识论，使之适于人类语言、逻辑、科学和学习等方面的更为宽泛的理论。我们可以尝试构建个人知识与群体知识的数学模型，以便精确地刻画我们对知识形成条件的[115]粗糙直觉。我们可以比较不同历史时期和不同文化中产生的既有哲学理论的优势。其中有的关于知识的哲学主张被证明是混乱或自毁性的，但也有一些成果，譬如说知识与真理的特殊关联，经受住了时间的考验。如果我们还无法提前预知何种方法最适合产出关于知识本质的深层洞见，那么这部分地是由于我们仍然没有完全理解究竟什么是知识。但是，即使我们还没有充分了解知识的本质，我们现在也已经在这个古老的问题上取得了更好的进展。[116]

索 引

（条目后的数字为原书页码，见本书边码）

A

Academic Scepticism 学院派怀疑论 13—15, 100
analysis of knowledge 知识的分析 46—59
Arcesilaus 阿凯西劳斯 13
Arnauld, Antoine 安东尼·阿诺德 38

B

BonJour, Laurence 劳伦斯·邦茹 68—70
brain-in-a-vat scenario 钵中之脑图景 22—28

C

Carneades 卡尼底斯 13
Cartesian Circle 笛卡尔主义学派 38
causal theory of knowledge 知识的因果理论 50—53
Chalmers, David 戴维·查尔莫斯 26—28
classical analysis of knowledge 知识的经典分析 48, 102
Closure Principle for knowledge 知识的封闭原则 91—92
Copernicus, Nicolaus 尼古拉·哥白尼 31
common knowledge 共有知识 3, 72, 83—85
confidence 信心 9, 101
contextualism 语境主义 87—96, 98, 100, 112
Craig, Edward 爱德华·克雷格 85—86
criterion of truth 真理标准 15
Cummins, Robert 罗伯特·卡明斯 114
Cynical Theory 犬儒主义理论 5—6

D

Descartes, René 勒内·笛卡尔 17—18, 33—39, 44
Dharmottara 法上 58
direct theory of testimony 证言知识的直接理论 80—81, 84—85
dogmatism 独断论 16
donkeys, cleverly disguised 巧妙伪装过的驴子 89—92, 108
dreams 梦境 12, 19—20, 38, 88

E

Early Modern period 近代早期 29—45
egocentric bias 自我中心的偏见 107—108
Elizabeth of Bohemia, Princess 波希米亚的伊丽莎白公主 38
empiricism 经验主义 39—44
evil demon scenario 恶魔图景 17—18, 22, 33—34
experimental philosophy 实验哲学 109—113
externalism 外在论 45, 60—67, 69—71, 93

F

factivity 事实性 8
facts 事实 2–3, 9
Fake Barn County 假谷仓县 52–53, 65–66
false belief task 虚假信念任务 104, 111

G

Galilei, Galileo 伽利略・伽利雷 30–31
Gaṅgeśa 甘吉沙 58
Gautama, Akṣapāda 阿克沙巴德・乔达摩 80–81
Generality Problem 普遍性问题 65–67, 70
Gettier, Edmund 埃德蒙德・盖梯尔 48–49
Gettier cases 盖梯尔反例 46–53, 58, 85–86, 102, 110–112
God 上帝 3, 35–38
Goldman, Alvin 阿尔文・古德曼 51–53, 65–66
group knowledge 群体知识 3

H

Hermann Grid illusion 赫尔曼栅格错觉 113
Hobbes, Thomas 托马斯・霍布斯 38

I

Inference to the Best Explanation 最佳解释推理 21–23
innate ideas 天赋观念 35–41
interest-relative invariantism 利害相关的不变论 97–99
internalism 内在论 60–62, 66–71, 93
intuitions about knowing 知识直觉 101–116

J

justification 辩护 48–49, 55, 61, 67–68
Justified True Belief analysis of knowledge “得到辩护的真信念”知识定义 48

K

knowledge-first epistemology 知识优先的知识论 56–57, 59, 104–105

L

Lackey, Jennifer 詹妮弗・拉基 82–83
linguistic features of ‘know’ “知道”的语言学特征 6–7, 105
Locke, John 约翰・洛克 38–44, 72–77
lotteries 彩票 46, 54

M

Matrix, The《黑客帝国》25–26
memory 记忆 51–52, 60, 76
mindreading 读心 103–109
moderate invariantism 温和的不变论 101, 113

Montaigne, Michel de 蒙田 17, 33
Moore, G. E. G. E. 摩尔 18—21

N

Nozick, Robert 罗伯特・诺齐克 62—65

O

objectivity 客观性 9—11

P

Paracelsus 帕拉塞尔苏斯 30, 32
Plato 柏拉图 11, 13, 81—82
power 权力 5
Protagoras 普罗塔哥拉 10—11
Putnam, Hilary 希拉里・普特南 24—25
Pyrrho of Elis 埃利斯的皮罗 13, 15—16
Pyrrhonian Scepticism 皮罗主义怀疑论 13, 15—16

R

rationalism 理性主义 33—38, 44
reductionist theory of testimony 证言知识的还原论 77—80, 83—84
relativism 相对主义 10—11
Relevant Alternatives theory of knowledge 知识的相关选项理论 89—92
reliabilism 可靠主义 53—54, 65—66
Russell, Bertrand 伯特兰・罗素 20—22, 46

S

Saxe, Rebecca 丽贝卡・萨克斯 105—106
scepticism 怀疑论 12—29, 43, 54, 88, 93, 94, 100, 112
Semantic Externalism 语义外在论 23—28
Sextus Empiricus 恩披里柯 15—16
Sizzi, Francesco 弗朗西斯科・希兹 31
Śrīharśa 室利曷沙 17
Stine, Gail 盖尔・斯泰恩 92
Stoic theory of knowledge 斯多葛主义知识论 13—14, 29, 36
strict invariantism 严格的不变论 100—101

T

testimony 证言 51—52, 72—86
tracking theory of knowledge 知识的追溯理论 62—65
truth 真理 7—11, 35, 116

U

universality of 'know' "知道"的普遍性 6—7

V

virtual reality 虚拟现实 25—28
Vogel, Jonathan 乔纳森・沃格尔 23

W

Weinberg, Jonathan 乔纳森·温伯格 109—111

Weiner, Matt 马特·维纳 56

Wikipedia 维基百科 72, 83—85

Williamson, Timothy 蒂莫西·威廉姆森 29, 57

Z

Zagzebski, Linda 琳达·扎克泽博斯基 55—56

Jennifer Nagel

KNOWLEDGE

A Very Short Introduction

Contents

Acknowledgements i

List of illustrations iii

1 Introduction 1

2 Scepticism 12

3 Rationalism and empiricism 30

4 The analysis of knowledge 46

5 Internalism and externalism 60

6 Testimony 72

7 Shifting standards? 87

8 Knowing about knowing 102

References 117

Further reading 127

Acknowledgements

For helpful comments and discussion I am grateful to Elena Derksen, Emma Ma, Alex Tenenbaum, Sergio Tenenbaum, and Tim Williamson. I am also grateful to the Social Sciences and Humanities Research Council of Canada for their support.

List of illustrations

1 Everything known must be linked to a knower **4**
Shutterstock

2 What do you know? **13**
Charles Barsotti, The *New Yorker* Collection/The Cartoon Bank

3 The Paracelsian view of our relationship to the universe **32**

4 René Descartes, 1596–1650 **34**
Shutterstock

5 John Locke, 1632–1704 **40**
Library of Congress, reproduction number LC-USZ62-59655

6 A stopped clock **47**
Copyright: Peter Kurdulija

7 The suffering of past generations **84**
Roz Chast, The *New Yorker* Collection/The Cartoon Bank

8 A man stands with one of Marah Zoo's world famous painted donkeys, in Gaza City, Palestinian Territories, December 2009 **90**
Anastasia Taylor-Lind/VII

9 The Hermann Grid illusion **113**

10 Can we make progress? **115**
Charles Barsotti, The *New Yorker* Collection/The Cartoon Bank

Chapter 1
Introduction

Searching for knowledge

The hunt for knowledge has never been easier. Hard questions can be answered with a few keystrokes. Our individual powers of memory, perception, and reasoning can be double-checked by distant friends and experts, with minimal effort. Past generations would marvel at the number of books within our reach.

These new advantages don't always protect us from an old problem: if knowledge is easy to get, so is mere opinion, and it can be hard to spot the difference. A website that looks trustworthy can be biased, world-renowned authorities can follow misleading evidence down the wrong track, and illusions can distort what we ourselves seem to see or remember. What at first seemed like knowledge can turn out to be something less than the real thing. Reflecting on the difficulty of enquiry, we can find ourselves wondering exactly what this real thing might be. What is knowledge? What is the difference between just thinking that something is true and actually knowing that it is? How are we able to know anything at all?

These questions are ancient ones, and the branch of philosophy dedicated to answering them—epistemology—has been active for thousands of years. Some core questions have remained constant throughout this time: How does knowledge relate to truth? Do senses like vision and hearing supply us with knowledge in the same way as abstract reasoning does? Do you need to be able to justify a claim in order to count as knowing it? Other concerns have emerged only recently, in light of new discoveries about humanity, language, and the mind. Is the contrast between knowledge and mere opinion universal to all cultures? In ordinary language, does the word 'know' always stand for the same thing, or does it refer to something heavier in a court of law, and something lighter in a casual bus-stop conversation? What natural instinctive impressions do we have about what others know, and how much can these impressions tell us about knowledge itself?

Over the centuries, philosophers investigating knowledge have unearthed some strange puzzles and paradoxes. Philosophers have also developed some innovative solutions to these problems. After looking back at major historical developments in the theory of knowledge, this book presses forward into its central current debates. We begin with a tour of some features of knowledge that easily spark philosophical curiosity.

Knowledge and the knower

Knowledge is sometimes portrayed as a free-flowing impersonal resource: knowledge is said to be stored in databases and libraries, and exchanged through 'the knowledge economy', as information-driven commerce is sometimes called. Like many resources, knowledge can be acquired, used for diverse purposes, and lost—sometimes at great expense. But knowledge has a closer connection to us than resources like water or gold. Gold would continue to exist even if sentient life were wiped out in a catastrophe; the continued existence of knowledge, on the other hand, depends on the existence of someone who knows.

It's tempting to identify knowledge with facts, but not every fact is an item of knowledge. Imagine shaking a sealed cardboard box containing a single coin. As you put the box down, the coin inside the box has landed either heads or tails: let's say that's a fact. But as long as no one looks into the box, this fact remains unknown; it is not yet within the realm of knowledge. Nor do facts become knowledge simply by being written down. If you write the sentence 'The coin has landed heads' on one slip of paper and 'The coin has landed tails' on another, then you will have written down a fact on one of the slips, but you still won't have gained knowledge of the outcome of the coin toss. Knowledge demands some kind of access to a fact on the part of some living subject. Without a mind to access it, whatever is stored in libraries and databases won't be knowledge, but just ink marks and electronic traces. In any given case of knowledge, this access may or may not be unique to an individual: the same fact may be known by one person and not by others. Common knowledge might be shared by many people, but there is no knowledge that dangles unattached to any subject. Unlike water or gold, knowledge always belongs to someone.

More precisely, we should say that knowledge always belongs to some individual or group: the knowledge of a group may go beyond the knowledge of its individual members. There are times when a group counts as knowing a fact just because this fact is known to every member of the group ('The orchestra knows that the concert starts at 8 pm'). But we can also say that the orchestra knows how to play Beethoven's entire Ninth Symphony, even if individual members know just their own parts. Or we can say that a rogue nation knows how to launch a nuclear missile even if there is no single individual of that nation who knows even half of what is needed to manage the launch. Groups can combine the knowledge of their members in remarkably productive (or destructive) ways.

Is there knowledge beyond the knowledge of human individuals and groups? What should we say about what is known by non-human animals? Or by God, if there is a God? These questions

1. Everything known must be linked to a knower

threaten to pull us into difficult biological and theological debates. For this reason, most epistemologists start with the simpler case of the knowledge of a single human being (such as yourself). This kind of knowledge will be the main focus of this book. Knowledge, in the sense that matters here, is a link between a person and a fact. It remains challenging to describe this link, even when we restrict our attention to one knowing person and one known fact. What is it for you to know something, rather than merely believe it?

Spotting the difference

As soon as we ask about the difference between knowing something and just believing it, we might wonder how we can be so confident that there really *is* a difference. Consider the view that there is no real difference between knowledge and opinion. What if 'knowledge' is nothing more than a label we apply to the attitudes of the elite? In our culture, perhaps a Nobel Laureate's scientific research or a CEO's thoughts about his industry; in other times and places, the teachings of the high priest or the tribal elders. Across the board, the thoughts of underdogs get brushed off as superstitions and misconceptions. On this view—call it the Cynical Theory—whether someone's idea counts as knowledge or mere opinion would be determined by his status as a leader or a loser, and not by anything in the idea itself or its relation to reality.

It's not entirely crazy to think of knowledge as a status marker. It's surely right that 'knowledge' is an attractive label; to describe an attitude as knowledge is to rank it above many other attitudes. It is also plausible that there are strong links running both ways between knowledge and power: power typically delivers advantages that can help a person gain knowledge, and knowledge can often help a person gain power. It may even be true that our judgements about knowledge are often biased by the social standing of those we are evaluating. But the Cynical Theory says something more than that power and knowledge often go together, or are widely thought to go together: it says that there is nothing more to knowledge than the perception of power.

As a theory about how we actually use the term 'knowledge', the Cynical Theory fails to capture some relevant facts. It underestimates our ability to resist the ideas of the powerful: even winners of the Nobel Prize can be doubted and challenged. More generally, even if we are more likely to see the powerful as knowing, we can still recognize a distinction between knowing and just seeming to know: we are all aware of cases in which experts once widely thought to know something have been proven wrong. The Cynical Theory also misses something notable about the way we talk about knowledge in everyday life. The verb 'to know' is not reserved for our descriptions of top experts and leaders: it's one of the ten most common verbs in English. It's the default verb for describing what happens as a result of ordinary cases of seeing, hearing, and remembering things: you know what you had for dinner last night, who won the last US presidential election, and whether you are now wearing shoes.

English is not unusual: across a vast array of languages—Russian, Mandarin, Welsh, Spanish—a word meaning 'know' is among the most common verbs. English does have one confusing feature, shared with a number of other languages: our verb 'know' has two distinct senses in common use. In the first, it can take either a propositional complement or 'that'-clause (as in 'He knows that the car was stolen') or an embedded question (as in 'She knows who stole the car' or 'She knows when the theft occurred'). In the second, it takes a direct object ('He knows Barack Obama'; 'She knows London'). Many other languages use different words for these two meanings (like the French 'savoir' and 'connaître'). In what follows, we will be focusing on the first sense of 'knows', the kind of knowledge that links a person with a fact.

This sense of 'knows' has an interesting feature: there's a word for it in all of the world's 6,000+ human languages ('think' shares this feature). This status is surprisingly rare: an educated person has a vocabulary of about 20,000 words, but fewer than 100 of these are thought to have precise translations in every other language. Common words you might expect to find in every language (like 'eat' and 'drink') don't always have equivalents. (Some aboriginal languages in Australia and Papua New Guinea get by with a single word meaning 'ingest'.) Elsewhere, other languages make finer rather than rougher distinctions: many languages lack a single word translating 'go', because they use distinct verbs for self-propelled motions like walking and for vehicular motion. Sometimes the lines

Box 1 Words that do and do not appear in all languages

Universal	Not universal
know, think, see, want, hear, say	eat, drink, stop, hit, go, sit
I, you	she, he, they
good, bad	happy, sad
not, maybe, because, if, true, before, after	plant, tree, animal, bird, cold, hot

are just drawn in different places: where the common English pronouns 'he' and 'she' force a choice of gender, other languages have third-person pronouns that distinguish between present and absent persons, but not between male and female. Human languages have remarkable diversity. But despite this diversity, a few terms appear in all known languages, perhaps because their meanings are crucial to the way language works, or because they express some vital aspect of human experience. These universals include 'because', 'if', 'good', 'bad', 'live', 'die'... and 'know' (see Box 1).

Knowing vs thinking

What do we ordinarily do with this vital verb, and how is 'know' different from the contrasting verb 'think'? Everyday usage provides some clues. Consider the following two sentences:

> Jill knows that her door is locked.
> Bill thinks that his door is locked.

We immediately register a difference between Jill and Bill—but what is it? One factor that comes to mind has to do with the truth of the embedded claim about the door. If Bill just thinks that his door is locked, perhaps this is because Bill's door is not really locked. Maybe he didn't turn the key far enough this morning as he was leaving home. Jill's door, however, must be locked for the sentence about her to be true: you can't ordinarily say, 'Jill knows that her door is locked, but her door isn't locked.' Knowledge links a subject to a truth. This feature of 'knowing that' is called *factivity*: we can know only facts, or true propositions. 'To know that' is not the only factive construction: others include 'to realize that', 'to see that', 'to remember that', 'to prove that'. You can realize that your lottery ticket has won only if it really has won. One of the special features of 'know' is that it is the most general such verb, standing for the deeper state that remembering, realizing, and the rest all have in common. Seeing that the barn is on fire or proving that there is no greatest prime number are just two of the many ways of achieving knowledge.

Of course, it's possible to *seem* to know something that later turns out to be false—but as soon as we recognize the falsity, we have to retract the claim that it was ever known. ('We thought he knew that, but it turned out he was wrong and didn't know.') To complicate matters, it can be hard to tell whether someone knows something or just seems to know it. This doesn't erase the distinction between knowing and seeming to know. In a market flooded with imitations it can be hard to tell a real diamond from a fake, but the practical difficulty of identifying the genuine article shouldn't make us think there is no difference out there: real diamonds have a special essence—a special structure of carbon atoms—not shared by lookalikes.

The dedicated link to truth is part of the essence of knowledge. We speak of 'knowing' falsehoods when we are speaking in a non-literal way (just as we can use a word like 'delicious' sarcastically, describing things that taste awful). Emphasis—in italics or pitch—is one sign of non-literal use. 'That cabbage soup smells *delicious*, right?' 'I *knew* I had been picked for the team.

But it turned out I wasn't.' This use of 'knows' has been called the 'projected' use: the speaker is projecting herself into a past frame of mind, recalling a moment when it seemed to her that she knew. The emphasis is a clue that the speaker is distancing herself from that frame of mind: she didn't literally or really know (as our emphatic speaker didn't really like the soup). The literal use of 'know' can't mix with falsehood in this way.

By contrast, belief can easily link a subject to a false proposition: it's perfectly acceptable to say, 'Bill thinks that his door is locked, but it isn't.' The verb 'think' is *non-factive.* (Other non-factive verbs include 'hope', 'suspect', 'doubt', and 'say'—you can certainly say that your door is locked when it isn't.) Opinions being non-factive does not mean that opinion is always wrong: when Bill just thinks that his door is locked, he could be right. Perhaps Bill's somewhat unreliable room-mate Bob occasionally forgets to lock the door. If Bill isn't entirely sure that his door is locked, then he could think that it is locked, and be right, but fail to know that it is locked. Confidence matters to knowledge.

Knowledge has still further requirements, beyond truth and confidence. Someone who is very confident but for the wrong reasons would also fail to have knowledge. A father whose daughter is charged with a crime might feel utterly certain that she is innocent. But if his confidence has a basis in emotion rather than evidence (suppose he's deliberately avoiding looking at any facts about the case), then even if he is right that his daughter is innocent, he may not really know that she is. But if a confidently held true belief is not enough for knowledge, what more needs to be added? This question turns out to be surprisingly difficult—indeed, difficult enough to warrant a whole chapter of its own (Chapter 4).

Because truth is such an important feature in the essence of knowledge, something further should be said about it here. We'll assume in what follows that truth is objective, or based in reality and the same for all of us. Most philosophers agree about the objectivity of truth, but there are some rebels who have thought otherwise. The Ancient Greek philosopher Protagoras (5th century BCE) held that knowledge is always of the true, but also that different things could be true for different people. Standing outdoors on a breezy summer day and feeling a bit sick, I could know that the wind is cold, while you know that it is warm. Protagoras didn't just mean that I know that the wind feels cold to me, while you know that it feels warm to you—the notion that different people have different feelings is something that can be embraced by advocates of the mainstream view according to which truth is the same for everyone. (It could be a plain objective fact that the warm wind feels cold to a sick person.) Protagoras says something more radical: it is true for me that the wind really *is* cold and true for you that the wind *is* warm. In fact, Protagoras always understands truth as relative to a subject: some things are true-for-you; other things are true-for-your-best-friend or true-for-your-worst-enemy, but nothing is simply *true*.

Protagoras's relativist theory of knowledge is intriguing, but hard to swallow, and perhaps even self-refuting. If things really are for each person as he sees them, then no one ever makes a mistake. It's true for the hallucinating desert traveller that there really is an oasis ahead; it's true for the person who makes an arithmetical error that seven and five add up to eleven. What if it later seems to you that you have made a mistake? If things always are as they appear, then it is true for you that you have made a mistake, even though appearances can never be misleading, so it should have been impossible for you to get things wrong in the first place. This is awkward. One Ancient Greek tactic for handling this problem involved a division of you-at-this-moment from you-a-moment-ago. Things are actually only true for you-right-now, and different things might be true-for-you-later (for example, it might be true-for-your-future-self that your past self made a mistake).

Splintering the self into momentary fragments is arguably a high price to pay for a theory of knowledge. If you find the price too high, and want to hold on to the idea that there is a lasting self, then Protagoras's theory may start to seem false to you. But if Protagoras's theory seems false, remember that by the lights of this theory you can't be mistaken: the theory itself tells you that things always are for you as they seem to you. Now Protagoras's theory is really in trouble. The self-destructive potential of relativism was remarked upon by Plato (*c.*428–348 BCE), who also noticed a tension between what Protagoras was trying to do in the general formulation of his theory, and what the theory says about truth being local to the individual. If Protagoras wants his theory of what is true-for-each-person-at-an-instant to capture what is going on with all of us, over time, it's not clear how he can manage that.

Relative truth has had more sophisticated defenders since Protagoras, but most philosophers have favoured objective truth. What is true is true for all of us, full stop, whether or not we are aware of it. If we have to put it in terms of perspective, what is true is what would be the case from a God's-eye perspective. But is objective truth humanly knowable? Sceptics have raised some doubts about this.

Chapter 2
Scepticism

Can you be sure?

Think of one of the most trivial and easily checked facts you know. For example, you know whether you are presently wearing shoes. Right?

The sceptic would like you to reconsider. Could you be dreaming that you are reading this book right now? If this is a dream, you could be lying barefoot in bed. Or you could be asleep on the commuter train, fully dressed. You may consider it unlikely that you are now dreaming, but you might wonder whether you have any way of establishing conclusively that you are awake, and that things are as they seem. Perhaps you remember reading somewhere that pinching yourself can end a dream—but did you read that in a trustworthy source? Or is it even something that you *really* read, as opposed to something that you are just now dreaming you once read? If you can't find any sure way of proving that you are now awake, can you really take your sensory experience at face value?

When you start to get self-conscious about what you know, even the simplest fact, something you usually think you could verify at a glance, can start to seem like something you don't really know. Evidence that you would usually take for granted suddenly seems dubious, and you may start to feel that certainty is slipping out of

2. What do you know?

reach, across the board. This pattern of thoughts has worried many philosophers over the centuries, deeply enough to drive some to doubt that human beings can have any substantive knowledge at all. These philosophers are 'sceptics', from the Ancient Greek for 'inquiring' or 'reflective'.

The historical roots of scepticism

Ancient Greece was in fact the birthplace of two distinct sceptical traditions, the Academic and the Pyrrhonian. Academic sceptics argued for the conclusion that knowledge was impossible; Pyrrhonian sceptics aimed to reach no conclusions at all, suspending judgement on all questions, even the question of the possibility of knowledge.

Academic Scepticism is named for the institution it sprang from: the Academy in Athens, originally founded by Plato. The movement's two great leaders each served a turn as head of the Academy: Arcesilaus in the 3rd century BCE, and then Carneades a hundred years later. Although both of them originally framed their scepticism in opposition to the once-influential Stoic theory of knowledge, their arguments continue to be taken seriously in philosophy to the present day. These sceptical arguments have enduring power because the core Stoic ideas they criticize are still embraced within many other theories of knowledge, and may even be part of our common-sense way of thinking about the difference between knowledge and mere belief.

Stoic epistemology draws a distinction between impressions and judgements. The Stoics noticed that you can have an impression—say, of shimmering water on a desert road—without judging that things really are as they seem. Judgement is the acceptance (or rejection) of an impression; knowledge is wise judgement, or the acceptance of just the right impressions. In the Stoic view, people make mistakes and fall short of knowledge when they accept poor impressions—say, when you judge that some friend is approaching, on the basis of a hazy glimpse of someone in the distance. When an impression is unclear, you might be wrong—or even if you are right, you are just lucky to be hitting the mark. A lucky guess does not amount to knowledge. The wise person would wait until that friend was closer and could be seen clearly. Indeed, according to the Stoics, you attain knowledge only when you accept an impression that is so clear and distinct that you couldn't be mistaken.

Academic Sceptics were happy to agree that knowledge would consist in accepting only impressions that couldn't be wrong, but they proceeded to argue that there simply are no such impressions. Would it be enough to wait until your friend comes nearer? Remember that people can have identical twins, with features so similar that you can't tell them apart, even close up. If you feel sure your friend has no twin, remind yourself that you might be misremembering, dreaming, drunk, or hallucinating. If the wise person waits to accept just impressions that couldn't possibly be wrong, he will be waiting forever: even the sharpest and most vivid impression might be mistaken. Because impressions are always fallible, the Academic Sceptics argued, knowledge is impossible.

One might wonder about the internal consistency of this position: how could the Academics be so certain of the impossibility of knowledge while at the same time doubting our ability to establish anything with certainty? Such concerns helped to motivate an even deeper form of scepticism. Imagine a way of thinking which consists of *pure* doubt, making no positive claims at all, not even the claim that knowledge could be proven to lie out of reach. Pyrrhonian Scepticism aimed to take this more radical approach. The movement was named in honour of Pyrrho of Elis (*c*.360–270 BCE), who is known to us not through his own written texts—he left no surviving works—but through the reports of other philosophers and historians. As a young man, Pyrrho joined Alexander the Great's expedition to India, where he is said to have enjoyed some exposure to Indian philosophy. On his return, Pyrrho started to attract followers, eventually becoming so popular that his home town honoured him with a statue and a proclamation that all philosophers could live there tax free. Pyrrho's influence now reaches us mainly through the writings of his admirer Sextus Empiricus (*c*.160–210 CE), who drew sceptical ideas from a range of ancient sources to form the branch of scepticism now known as Pyrrhonism.

A central worry of both schools of ancient scepticism concerns the 'criterion of truth' or the rule we should use to figure out what to accept, assuming that knowledge requires not just accepting things randomly. Different philosophers have proposed different rules for deciding which impressions are the best ones, and if knowledge demands thoughtful choice, our criterion of truth itself can't just be chosen randomly. But if we can't be arbitrary, we need some rule to select a criterion of truth. Do we use our favoured criterion to justify itself? That looks like circular reasoning. But if we look for a fresh rule to justify our criterion, then we will need some further principle to justify that fresh rule, and so on into an infinite regress.

Pyrrhonians were happy to draw attention to the difficulty of the problem of the criterion, without positively claiming that it would forever remain unsolved. But having appreciated the difficulty here, they adopted the plan of general suspension of judgement. They developed a general strategy for generating doubt on any topic at all: whenever you are tempted to make up your mind one way on an issue, consider the other way. Instead of settling the matter one way or another (which would be 'dogmatism'), just continue the search for further evidence, keeping the two sides of a question balanced against each other in your mind. Many techniques were available for keeping contrary ideas in balance. You could think about how different things would appear to other animals, or when seen from other perspectives, or in different cultures. Drawing on the work of earlier sceptics, Sextus Empiricus developed an extensive catalogue of ways to keep yourself from settling for any particular answer to any given question. He also developed lists of phrases that the sceptic could say to himself ('I determine nothing'; 'Perhaps it is, and perhaps it is not'). Sextus did not want these phrases to be seen as his own expressions of dogma: his scepticism was laid out as a practice or way of life, and not as a positive theory of reality. Keeping all questions open may sound like a recipe for anxiety, but curiously enough, Sextus reported the impression that his sceptical practice seemed to bring peace of mind. (Only an impression, of course—he couldn't be sure that he had achieved true peace of mind, or that it had come as a result of the scepticism, rather than by chance.)

One early criticism of scepticism was that it would be problematic for human survival: if sceptics suspend judgement even on the question of whether eating will satisfy their hunger, aren't they at risk of starvation? The Pyrrhonians suggested that behaviour can be guided by instinct, habit, and custom rather than judgement or knowledge: in resisting dogma, sceptics do not have to fight against their raw impulses or involuntary impressions. Sceptics can satisfy their hunger and thirst on autopilot while refraining from judgement about reality.

The sceptical path of resisting all judgement is not an easy one to follow, and throughout the Middle Ages the dominant figures of Western philosophy were firmly non-sceptical. Scepticism did flourish in the Indian tradition, however, most remarkably in the work of Śrīharśa, whose 11th century text *The Sweets of Refutation* promised to teach its readers some spectacular techniques of argument that could be used against any positive theory whatsoever. Śrīharśa promised his readers that they could attain 'the joy of universal conquest' in their arguments by using these sceptical methods, many of which focused on the difficulty of defining the terms we use. Like the Ancient Greeks, Śrīharśa made much of the deceptiveness of appearances, and the weakness of our powers to discover the true nature of things. Śrīharśa also seems to have shared Sextus's hope of enjoying peace of mind upon abandoning all positive attempts to attain knowledge.

Radical doubts about the possibility of knowledge emerged periodically over the following centuries, most strikingly during periods of intellectual upheaval. As new scientific work challenged the medieval world view in the 16th century, there was a resurgence of scepticism in Europe. The works of Sextus Empiricus were rediscovered, and his arguments were eagerly taken up by philosophers such as Michel de Montaigne (1533–92), whose personal motto ('What do I know?') expressed his enthusiasm for Pyrrhonian doubt. This sceptical spirit was contagious: early in the 17th century, René Descartes (1596–1650) reported that, far from being an extinct ancient view, scepticism was 'vigorously alive today'. Descartes's best-known work, *Meditations on First Philosophy*, presents truly novel sceptical arguments about the limits of reason, alongside familiar ancient arguments about dreaming and illusions. In his deepest sceptical moment, Descartes invites you to contemplate a scenario in which a powerful evil demon is dedicated to deceiving you at every turn, not only sending you illusory sensory impressions, but also leading you astray each time you attempt to make an abstract judgement such as a simple arithmetical calculation. This vivid scenario has lingered in the philosophical imagination, even though Descartes himself thought that there was a sure way to dispel it. Descartes was not himself a sceptic (despite his considerable talent for presenting sceptical arguments): he took himself to have an airtight proof that scepticism is mistaken. His optimism about having solved the problem of scepticism was not widely shared, however, and the major thinkers of the Early Modern period continued to struggle with the problem. Chapter 3 will examine Descartes and his rivals more closely, looking at their treatments of scepticism in the context of their positive theories of knowledge.

Old challenges, fresh replies

The old question of scepticism received some surprising new answers in the 20th century. A strangely simple approach was advanced by the English philosopher G. E. Moore in a public lecture in 1939. In answer to the question of how we could prove the reality of the external world, Moore simply held up his hands (saying, 'Here is one hand, and here is another'), explained that they were external objects, and drew the logical conclusion that external objects actually exist. Moore considered this to be a fully satisfactory proof: from the premise that he had hands, and the further premise that his hands were external objects (or, as he elaborated, 'things to be met with in space'), it clearly does follow that external things exist. The sceptic might, of course, complain that Moore did not really *know* that he had hands—but here Moore proposed shifting the burden of proof over to the sceptic. 'How absurd it would be to suggest that I did not really know it, but only believed it, and that perhaps it was not the case!' Moore insists that he knows that he has hands, but doesn't even try to prove that he is right about this. After shrugging off the sceptic's worries as absurd, Moore aims to explain why he won't produce a proof that he has hands, and why we should still accept him as having knowledge on this point.

Moore starts by remarking that when he claims to know (without proof) that he has hands, he is not claiming that a person can *never* prove he has hands. Moore was willing to grant that there are special situations in which someone might reasonably prove the existence of his hands: for example, if anyone suspects that you are an amputee with artificial limbs (and you are not), you could let him examine your hands more closely to dispel that particular doubt. If you were really keen to prove your point, you could even let him feel your pulse or scratch you with a sharp object. But however well that strategy might work to dispel particular doubts about artificial limbs, Moore does not think that there is an all-purpose strategy for proving that your hands exist, a general proof that would dispel all possible doubts. The range of possible doubts is truly enormous. To take just one example, a fully general proof against all doubts would have to show that you were not a sleeping amputee, dreaming in your hospital bed after an accident in which you lost your arms. Moore is pessimistic about anyone's chances of proving that this (admittedly far-fetched!) scenario isn't happening. However, just as Moore thinks you could know that you have hands without being able to prove as much, he also thinks that your inability to prove that you are not dreaming does not stop you from knowing that you are not dreaming. Moore once again retains confidence in his knowledge despite the limitations on what he is able to prove: 'I have, no doubt, conclusive reasons for asserting that I am not now dreaming; I have conclusive evidence that I am awake: but that is a very different thing from being able to prove it. I could not tell you what all my evidence is; and I should require to do this at least, in order to give you a proof.'

In claiming to have conclusive evidence that he is awake, Moore is resisting the hypothetical push of the sceptic's reasoning. Moore actually agrees with the sceptic that *if* you are dreaming, then you will not know just by looking that you have hands. But Moore reminds us that the sceptic's argument rests on that big 'if': as Moore sees it, the person who knows that he is not dreaming (whether or not he can prove this) should not be troubled by the sceptic's worries.

The strategy of declaring that one has knowledge without proof may set off some alarm bells (is Moore declaring victory after refusing to fight?). It may also seem odd that Moore is willing to construct what he thinks is a very good proof of the claim 'External objects exist,' while simply turning his back on the project of proving the somewhat similar claim 'These hands exist.' There's an important difference between those two assertions, however: the first is general and philosophical in character, and the second particular and ordinary. Explicit reasoning or proof has a clear role to play when we are supporting general philosophical claims: we can engage in extended reasoning about exactly what it means for something to be an 'external object', and indeed much of Moore's lecture is taken up with detailed discussion of this issue. By contrast, an ordinary claim like 'Here is a hand' is so basic that it is hard to find simpler and better-known claims we could use to support it. (There is a parallel with mathematics here, where some basic claims are taken as axioms, not themselves in need of proof.) If the sceptic attempts to undermine our certainty about such basic matters, Moore would urge us to distrust the sceptic's fancy philosophical reasoning well before we distrust our original common sense. We could reasonably be alarmed by someone claiming to know a controversial philosophical claim despite an inability to prove it; we should not feel such resistance to someone who claims to know a simple observable fact about his immediate environment.

Even philosophers who are receptive to Moore's suggestion that there is something wrong with the sceptic's reasoning may feel unsatisfied with Moore's plain and stubborn insistence on his common-sense knowledge. Some have tried to identify more precisely what mistake the sceptic is making, while also constructing a positive defence of our common-sense claims to knowledge. One major strategy was advanced by Bertrand Russell, a colleague of Moore's at Cambridge. Russell grants one point to the sceptic right away: it is *logically possible* that all of our impressions (or 'sense data', to use Russell's terminology) have their origin in something quite different from the real world we ordinarily take ourselves to inhabit. But in Russell's approach to scepticism, now known as the 'Inference to the Best Explanation' approach, we can grant that point about logical possibility and still hang on to fight the sceptic. Russell argues that there is a large gap between admitting that something is logically possible and concluding that we can't rationally rule it out: we have rational principles other than the rules of logic, narrowly conceived. In particular, Russell invokes the principle of simplicity: other things being equal, a simpler explanation is rationally preferred to a more complex one. It's logically possible that all the sense data you ordinarily credit to your pet cat (meowing sounds, the sight and feel of fur, and so on) do not come from the source you expect. Perhaps these impressions issue from a succession of different creatures, or from a series of inexplicably consistent dreams or some other strange source. But the simplest hypothesis, according to Russell, is the one that you would most naturally believe: there is a single real animal whose periodic interactions with you cause the relevant cat-like impressions in the stream of your private experience. Just as it is rational for scientists to explain patterns in their data by appeal to simple laws, it is rational to explain patterns in our everyday experience by appeal to a simple world of lasting objects (the 'real-world' hypothesis).

Russell's approach has its attractions, but a few worries may linger. Even if we grant that Inference to the Best Explanation is generally a rational strategy, we might feel that it seems insufficiently conclusive to ground *knowledge* as opposed to just rational belief. This potential weakness of Inference to the Best Explanation can be illuminated by thinking about other contexts in which this style of reasoning is used. For example, a detective might use Inference to the Best Explanation when investigating a crime: once he has found mud matching the crime scene on the butler's shoes, heard the maid's testimony about the butler's hatred of the victim, and discovered a bloody knife under the butler's bed, the detective could reasonably conclude that the best explanation of the available evidence is that the butler committed the murder. However, the sceptic could point out, things might not be as they seem: perhaps the maid has committed the murder and very skilfully framed the innocent butler. This is not the simplest explanation, but it just might be the true one. Assuming that the detective uncovered no evidence of the maid's involvement, it could be rational for him to conclude that the butler is guilty, but this wouldn't establish that the butler actually was guilty. Likewise, some sceptics might be willing to grant that it is very likely that our experiences arise from ordinary external objects, or even that it is reasonable to believe as much, without being willing to grant that these experiences give us knowledge: knowledge, they might argue, calls for a higher standard than rational belief.

A further worry about Russell's strategy is that it is not obvious that the real-world hypothesis really is a better explanation of our experience than rival explanations the sceptic might offer. A sceptic might argue that the evil demon hypothesis can neatly explain the very features of our experience that impressed Russell: *of course* an evil demon would send us vivid and apparently coherent experiences over time, given that the evil demon is trying to deceive us into believing that there is an outer world of objects. This objection applies equally well to other versions of the sceptical hypothesis. For those who resist the supernatural element of the evil demon story, there is a modernized scientific version available: just suppose that your brain has been removed from your body and connected to a supercomputer which simulates experiences of a coherent reality, sending signals along your sensory pathways. If the program is good enough, maintaining consistency over time and adjusting its displays to match your outgoing motor signals (you decide to look to the left, and your visual experience changes accordingly...), your experience as a brain in a vat might be internally indistinguishable from the experience of someone interacting with an ordinary physical environment. Everything you think you see and feel—the blue sky outside, the warmth of the sun—could be an element in the large-scale virtual reality simulated by the supercomputer. Assuming that the point of the whole simulation is to give you sensory experiences that perfectly mirror the sensory experiences you'd have in an ordinary physical world, the challenge to the advocate of the Inference to the Best Explanation approach would be to explain why exactly the real-world hypothesis is a better explanation of our experience than the brain-in-a-vat hypothesis. To answer this challenge, various suggestions have been advanced: for example, the American philosopher Jonathan Vogel has argued that the basic spatial structure of the real world is much simpler than the spatial structure of the brain in a vat's virtual-reality-within-a-real-world, making the real-world hypothesis a better way to explain our experience.

Even if Vogel is right that it would be somewhat more reasonable to take ourselves to inhabit the real world rather than a vat, we may still crave a more powerful response to the sceptic. Is there a way to show that the sceptic's arguments contain a deep confusion, or that the sceptic must be saying something that is not just implausible but flat-out *wrong*? In recent years, some philosophers have used tools from the philosophy of language in an attempt to attack scepticism more aggressively. The motivating thought behind this new approach (the 'semantic approach') is that we can find ammunition against the sceptic by looking closely at the way in which our words have meaning or link up to reality. In particular, these new arguments against scepticism have drawn on a movement in the philosophy of language known as Semantic Externalism, a movement that traces back to the work of Ruth Barcan Marcus, Saul Kripke, and Hilary Putnam in the 1960s and 1970s.

The key idea of Semantic Externalism is that words get their meanings not from the images or descriptions that individual speakers associate with those words in their minds (that would be 'Semantic Internalism'), but from causal chains connecting us to things in the world around us. For example, in Shakespeare's time, water was thought to be an element; modern scientists now characterize it as the compound H_2O. But even if the question 'What is water?' would be answered quite differently by Shakespeare, the modern scientist, and the average person on the street, we have all been interacting with the same substance. The Semantic Externalist contends that we can all refer to the same substance when we say 'water' exactly because our meaningful use of the word is anchored in our common causal contact with a particular substance. Because Shakespeare and the modern scientist have seen and tasted the same liquid, whatever they thought about its nature, we can now say that they mean the same thing when they use the word 'water'. Sometimes the relevant causal chains must run through other speakers: no person alive today has met the late French emperor Napoleon, but we can still talk about him, as long as we pick up our use of the word 'Napoleon' through sources with the right kind of causal links back to the man himself. Semantic Externalism is especially useful in explaining how speakers with different (and conflicting) ideas about something can still talk about the same thing: when Jill says that Napoleon was very short, and Bill says that Napoleon was actually above average in height (knowing that the rumours of his small stature were started by his English enemies), they can be discussing the same person despite the differences in their mental images.

The best-known application of Semantic Externalism to scepticism is found in Hilary Putnam's 1981 book *Reason, Truth and History*. Putnam's target was the sceptic who attempts to make you feel very worried that you might be a brain in a vat. Putnam argued that the sentence 'I am now a brain in a vat' couldn't actually be true for anyone who understands that sentence. According to Putnam, a creature that only ever had electronic stimulation of a virtual reality could not mean what we do by the word 'vat': because this creature has only interacted with simulated images of vats, his word 'vat' can't refer to something physical in the world beyond the simulation. Outside the simulation, our grasp of the word's meaning is anchored in our history of experiences with real-world vats. Our very capacity to make sense of the sceptic's hypothesis is a sign that we are in the real world: if you actually understand what vats are, it can't be the case that you have always been a brain in a vat.

Putnam's argument inspired an immediate backlash. Some said that it only deepened the sceptical problem: now we should worry that perhaps our words don't even have the meaning that we thought they did. Others noticed that even if Putnam could show that someone who had always been a brain in a vat couldn't talk about real vats, never having had the right kind of experience with them, Putnam's idea wouldn't work for a brand new brain in a vat. The sceptic isn't restricted to trying to make you worry that you *are now and have always been* a brain in a vat. The creative sceptic could try instead to make you worried that you have been a brain in a vat just since midnight last night: you could suppose that your life until last night was perfectly normal, and then while you were sleeping a mad scientist took your brain and hooked it up to his supercomputer, which simulates a virtual world very like the reality you inhabited before last night. Assuming it's a very good simulation, nothing in your present experience can rule this out, and, unfortunately for Putnam, Semantic Externalism delivers the verdict that your word 'vat' is still perfectly meaningful, with its meaning anchored in your past experiences of real-world vats. This 'recent vat' scenario still has the result that all of your present sensory experience could be illusory (you think the sky is blue, but it's actually a dark and stormy day, and so on…), and this is a disturbing result that the sceptic will be delighted to secure.

The 1999 science fiction movie *The Matrix* dramatized this problem for a large audience. The movie's hero ('Neo', played by

Keanu Reeves) discovers that his life as an office worker in the 1990s is just a simulation: it's actually two difficult centuries later, and human beings have lost a war against machines guided by artificial intelligence. These machines now keep humans trapped in pods, and, for reasons that it would be tedious to explain, feed them with synthesized experiences in a massive synchronized virtual reality ('the Matrix'). By discovering some weaknesses in the programming, and by getting some help from a few brave humans who have resisted the machines, Neo finds a way out of the simulation. He is offered a choice between continuing life as a comfortable office worker in the simulation, and breaking free of the Matrix to join the underground fight against the machines, a moment dramatized in the movie as a choice between taking a red pill (for dangerous reality outside the Matrix) or a blue pill (for continued comfort within it). Fortunately for the plot of the movie, Neo takes the red pill. But one might wonder what is at stake here. Why exactly would experiencing a world of physical objects be better than staying immersed in a computer-generated simulation?

Perhaps it wouldn't. This surprising answer has been vigorously defended by the philosopher David Chalmers, who argues that the brain in a vat (BIV) is better off than you might suppose. In particular, Chalmers aims to show that the BIV's everyday beliefs about its surroundings are *true*: when the BIV is stimulated by the supercomputer to 'see' a book, for example, and then thinks to itself 'I am holding this small book in my hands,' it isn't actually being deceived or making a mistake. The trick is just that the BIV's words and thoughts refer not to objects composed of physical particles but to something fundamentally computational in character. Picking up on the Semantic Externalist idea that links the meanings of our words to whatever caused our experiences, Chalmers proposes that when the BIV talks about 'this small book' or 'my hands', it is actually referring to the subroutines in the supercomputer that are responsible for its relevant sensations of white paper and grasping fingers. This means the BIV is *right* to say that it is holding the book, given the nature of its experiences—the part of the supercomputer's program responsible for the book experiences is indeed going to be linked up to the part responsible for his hand experiences in a way that constitutes 'holding' for the BIV, the only kind of holding the BIV has ever known. From our perspective outside the vat, we'd say that the BIV is virtually holding a virtual book, but from within, the BIV is right to say and think that it is holding a book: its words have meanings appropriate to its environment, and given those meanings, what the BIV believes is true. In fact, nothing stops the BIV from *knowing* that it is holding (what it would call) a book. The 'sceptical scenario' doesn't really threaten everyday knowledge at all.

On this way of thinking, a virtual reality that is ultimately computational in character is no worse than a physical reality that is ultimately composed of subatomic particles. It might surprise us to discover that we inhabit such a reality, but if Chalmers is right, such a discovery would be like the discovery that quantum mechanics or string theory is true: we might be momentarily unsettled to hear that the ultimate nature of things is stranger than we had previously imagined, but the news should not shake our faith in our ordinary knowledge about shoes, hands, and books. (You care whether your shoes are wet or uncomfortable, but do you really care whether your shoes are ultimately composed of point-like particles, vibrating one-dimensional objects, or instructions in a computer program? Chalmers thinks: probably not.)

The sceptic wants to use the BIV scenario to scare us into thinking that there is no knowledge: 'You can't possibly rule out a situation in which most of your beliefs are false.' Unlike Moore, Chalmers agrees with the sceptic that the BIV scenario can't be ruled out, but he fights back against the suggestion that most of our beliefs would be false in such a situation. However, we might worry about variations on the sceptical scenario: like Putnam, Chalmers has difficulty with cases involving subjects very newly immersed in a virtual reality. Because they have known in the past what it is to physically hold a non-virtual book, and because the meanings of their words are anchored to their past experiences, they will be saying something false when their newly attached supercomputer gives them virtual reality experiences of a virtual book: when the fledgling BIV says 'I am holding a book,' it will mean a real, particle-based book, and what it says will therefore be false, and it will fail to have knowledge. Chalmers is very much aware of this worry, but seeks to soften its impact by pointing out that the newly envatted BIV will still have extensive knowledge of its past. If Chalmers is right, the sceptic can't easily generate a sweeping all-purpose scenario in which *almost all* of your beliefs about your past, present, and future are false. The persistent sceptic could still point out that his main knowledge-undermining purposes can be served pretty nicely by a milder scenario in which you are just (for example) wrong about everything you now perceive. If you've just been hooked up to a supercomputer that is simulating your experiences, then even Chalmers will admit that you can't know whether you are now wearing shoes. Meanwhile, you might continue to worry that your inability to rule out this scenario should undermine your claim to know such a fact already, whether or not you are in a simulation.

Is it hopeless to try to argue against the sceptic in the first place? In contrast to the ambitious project of attacking the sceptic's reasoning, some philosophers have advocated a more defensive approach, according to which we must resist the temptation to try to answer the sceptic on his own terms, and instead aim to stay out of his reach in the first place. Given that the sceptic has promised to raise doubts about everything you say, there is not much hope for finding common ground with him on the basis of which you will convince him that he is mistaken. Those who choose to start with only premises the sceptic accepts will have trouble digging themselves out of the pit. Those who choose instead to start from a common-sense world view have the easier task of building some defence mechanisms against the charms of scepticism, while accepting our common-sense assumption that there is a good deal we know.

One way to defend yourself against scepticism is to come up with a diagnosis of why it seems so appealing, despite its tendency to lead us towards strange or even absurd conclusions. Various proposals have been advanced. Oxford philosopher Timothy Williamson suggests that scepticism initially looks appealing (despite its bleak consequences) because it is a good thing carried too far. The good thing is that we have a healthy critical ability to double-check individual things we believe by suspending judgement in them temporarily to see whether they really fit with the rest of what we know. But if this ability to suspend individual beliefs serves as a useful immune system to weed out inconsistent and ungrounded ideas, scepticism is like an autoimmune disease in which the protective mechanism goes too far and attacks the healthy parts of the organism. Once we have suspended too much—for example, once we have brought into doubt the reality of the whole outer world—we no longer have the resources to reconfirm or support any of the perfectly reasonable things we believe.

Other factors may also play a role in the appeal of scepticism. Perhaps there is something wrong with the basic model of knowledge that forms the background of so much sceptical argument. On this model, inherited from Stoicism, we start with impressions (or ideas, or sense data), and then we have a distinct stage of accepting or rejecting them. If we agree that the impressions received in real life and dreams or simulations seem just the same to us, it's not obvious how we could ever do a good job of choosing which ones to accept. This question loomed especially large in philosophy during the Early Modern period.

Chapter 3
Rationalism and empiricism

The Early Modern period

In many fields—literature, music, architecture—the label 'Modern' stretches back to the early 20th century. Philosophy is odd in starting its Modern period almost 400 years earlier. This oddity is explained in large measure by a radical 16th century shift in our understanding of nature, a shift that also transformed our understanding of knowledge itself. On our Modern side of this line, thinkers as far back as Galileo Galilei (1564–1642) are engaged in research projects recognizably similar to our own. If we look back to the Pre-Modern era, we see something alien: this era features very different ways of thinking about how nature worked, and how it could be known.

To sample the strange flavour of pre-Modern thinking, try the following passage from the Renaissance thinker Paracelsus (1493–1541):

> The whole world surrounds man as a circle surrounds one point. From this it follows that all things are related to this one point, no differently from an apple seed which is surrounded and preserved by the fruit... Everything that astronomical theory has profoundly fathomed by studying the planetary aspects and the stars... can also be applied to the firmament of the body.

Thinkers in this tradition took the universe to revolve around humanity, and sought to gain knowledge of nature by finding parallels between us and the heavens, seeing reality as a symbolic work of art composed with us in mind (see Figure 3).

By the 16th century, the idea that everything revolved around and reflected humanity was in danger, threatened by a number of unsettling discoveries, not least the proposal, advanced by Nicolaus Copernicus (1473–1543), that the earth was not actually at the centre of the universe. The old tradition struggled against the rise of the new. Faced with the news that Galileo's telescopes had detected moons orbiting Jupiter, the traditionally minded scholar Francesco Sizzi argued that such observations were obviously mistaken. According to Sizzi, there could not possibly be more than seven 'roving planets' (or heavenly bodies other than the stars), given that there are seven holes in an animal's head (two eyes, two ears, two nostrils and a mouth), seven metals, and seven days in a week.

Sizzi didn't win that battle. It's not just that we agree with Galileo that there are more than seven things moving around in the solar system. More fundamentally, we have a different way of thinking about nature and knowledge. We no longer expect there to be any special human significance to natural facts ('Why seven planets as opposed to eight or 15?') and we think knowledge will be gained by systematic and open-minded observations of nature rather than the sorts of analogies and patterns to which Sizzi appeals. However, the transition into the Modern era was not an easy one. The pattern-oriented ways of thinking characteristic of pre-Modern thought naturally appeal to meaning-hungry creatures like us. These ways of thinking are found in a great variety of cultures: in classical Chinese thought, for example, the five traditional elements (wood, water, fire, earth, and metal) are matched up with the five senses in a similar correspondence between the inner and the outer. As a further attraction, pre-Modern views often fit more smoothly with our everyday

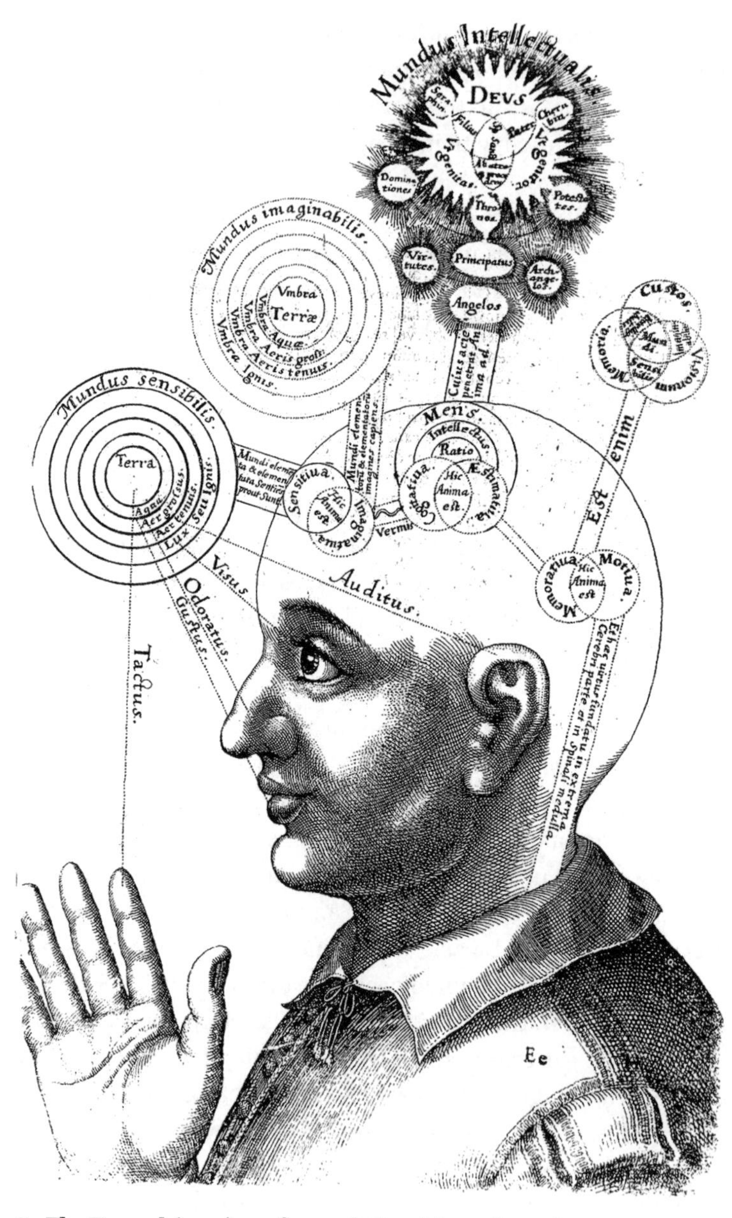

3. The Paracelsian view of our relationship to the universe

sense experience: naively, the earth looks to be stable and fixed while the sun moves across the sky, and it takes some serious discipline to convince oneself that the mathematically more simple models (like the sun-centred model of the solar system) are right.

For a brief period in the Renaissance the old and new ways of thinking fought for dominance. The conflict between these ways of thinking drove some philosophers to scepticism. Looking at the clash between the old theory that put the earth at the centre of everything and the new theory that the earth revolved around the sun, Montaigne decided that the only sensible path was not to take sides either way. But not all philosophers were happy to stay on the sidelines: some wanted to fight for the new, scientific way of thinking.

The Rationalism of René Descartes

Born in France in 1596, Descartes (see Figure 4) received a traditional Jesuit education, steeped in the classical ideas of Aristotle and his medieval interpreters. When he was later exposed to the newly emerging ways of studying nature, he had second thoughts about what he had learned in school. His *Meditations* begins with the confession that he had swallowed a 'large number of falsehoods' as a child. 'I realized,' he continues, 'that it was necessary, once in the course of my life, to demolish everything completely and start again right from the foundations if I wanted to establish anything at all in the sciences that was stable and likely to last.' The demolition programme that Descartes maps out in the *Meditations* was the method of systematic scepticism mentioned in Chapter 2, starting with small doubts about sensory illusions and finally rising to the all-consuming doubt generated by the evil demon scenario. By the time the evil demon appears at the end of the First Meditation, the narrator announces that he has succeeded in wiping the slate clean: he has generated doubts about all of his prior opinions.

4. René Descartes, 1596–1650

The remaining five Meditations carefully reconstruct a new world view on what Descartes hopes will be an unshakable basis. The first truth that can be known with utter certainty is the fact of one's own existence: whether or not an evil demon is tricking you at every step, you can never be mistaken about the fact that you yourself exist, and are a thinking being, at the moment when you think to yourself that you do exist. So far so good. But if 'I exist' seems like a slim basis to support the rest of what can be known, Descartes has an interesting trick for getting extra mileage out of it. In the Third Meditation he raises a reflective question: what was it, exactly, that enabled him to know that first fact? There was nothing more, he says, than 'a clear and distinct perception of what I was asserting'. If that is right, then we should be able to trust the rule *Whatever is clear and distinct is true*—for if this rule itself were fallible, we would not have a good basis to claim knowledge even of our own existence, Descartes reasons. The Third Meditation then bravely sets out to defend this general rule.

Without having established anything more than the fact of his own existence, Descartes has sharply limited resources to draw on, but as he continues to look within himself, he discovers that he has a variety of ideas in his mind, and he starts an inventory. Without yet deciding whether they represent anything real, Descartes notices that there are striking differences among his ideas. They differ in the kinds of things they represent (angels, men, the sun, the sky, God) and in their apparent origins. Some ideas (like his idea of the fireplace which is apparently warming him) seem to come from external things. Some ideas (like his idea of the hippogriff, a flying eagle-headed horse) seem to be invented, and others (like his idea of truth) seem to be built into him, or innate. As long as Descartes is worried that the evil demon might be deceiving him at every turn, it seems impossible for him to establish that any idea of his really does have the source it seems to have. But Descartes at last finds what seems like a way forward in the recognition that his different ideas represent greater and lesser things. Among all of his ideas, there is one that stands out to him at the top of the scale: his idea of God, an infinite and perfect being. The idea of perfection couldn't possibly come from any less-than-perfect source, Descartes reasons, and he concludes that this idea must come from God himself, and must be innate, planted within us from the start. Descartes's way of arguing from the mere idea of God to the real existence of God involves an appeal to a causal principle that not all of his readers have found wholly convincing (the principle that perfection, even in a mere idea, can never be caused by a less-than-perfect source). But once the existence of God is taken as established, Descartes moves swiftly to argue that all of our clear and distinct ideas (like our ideas of truth, numbers, and pure geometrical forms) are innate and entirely trustworthy. A perfect (and therefore benevolent) God would not have installed defective innate ideas in us.

If we have innate clear and distinct ideas that are perfectly trustworthy, it is an interesting question how we can ever make mistakes (as we obviously do). Descartes's next task is to tease apart the sources of error from the purely trustworthy core of our clear and distinct ideas. Some ideas, says Descartes, are obscure and confused rather than clear and distinct. There is some meaning in raw sensations like hunger, thirst, or even the sensations of colour and scent, but these ideas do not show us the true nature of things as they are in themselves. You can be wrong about the colour of an object in dim lighting, because colour ideas are not clear and distinct. By contrast, a clear and distinct idea like your idea of the number five, or the shape of a triangle, cannot fail to show you the nature of the thing it represents: you do not have to worry that your idea of the number five is misrepresenting that number. Echoing Stoic ways of thinking, Descartes contends that we are at risk of error when we make judgements on the basis of what is obscure and confused: as long as we stick to the clear and distinct we cannot fail. The best way to gain knowledge of something confused like scent or colour would be to analyse it in clear mathematical or geometrical terms (as we Moderns now do, explaining scents in terms of the shapes of molecules, and colours in terms of measurable wavelengths of light).

Known as rationalism, this way of thinking puts abstract concepts at the heart of our pursuit of knowledge. Descartes, whose own contributions to algebra and geometry were substantial (and included the development of Cartesian coordinates in geometry), was keen to defend the Modern enterprise of using mathematical tools to analyse nature. However, he still faced some difficulties in explaining the source of human error, an especially tricky project given his reliance on the claim that our innate intellectual hardware comes from a perfect God. In the Fourth Meditation, Descartes argues that God cannot be blamed for our mistakes, because these mistakes arise as a consequence of our freedom—in this case, human freedom to accept ideas that are less than clear and distinct. Freedom is, of course, not a bad thing, and any abuse of it is our fault, and not God's. It remains a mystery at this point, however, why God ever gives us obscure and confused bodily sensations alongside our brilliant and perfectly accurate mathematical and abstract ideas. It is only when he reaches the Sixth and final Meditation that Descartes at last tackles the difficult question of the point of our sensations.

Within the mind, Descartes suggests, we have various powers, including not only the intellect, which is responsible for abstract operations like those in arithmetic and deductive geometry, but also the imagination. To appreciate the difference in these powers, Descartes urges the reader to think of a chiliagon, a geometrical figure with exactly 1,000 sides. Intellectually, you can do this with perfect precision, appreciating the exact difference between a chiliagon and a 1,001-sided figure, or even calculating, if you remember the right formula, that a chiliagon has internal angles summing to exactly 1,996 right angles. But the imagination is rougher: when you try to visualize a chiliagon, Descartes predicts you will come up with something more confused, a figure with many, many sides, hard to tell apart from any other figure with many, many sides. Descartes concludes that the imagination is not simply reading the clear and distinct ideas in the intellect, but is oriented towards something else—the bodily senses.

The purpose of our sensations and related powers of imagination, Descartes ultimately concludes, is not to show our souls the true natures of things as they are in themselves (that job is reserved for the intellect); rather, sensations serve the interests of the body and soul taken together. Sensations like hunger, pain, scent, and colour help to ensure our bodily survival, which is itself something that a benevolent God could, of course, have wanted to protect. We can avoid being misled by our sensations if we keep in mind that they are not designed to show the true nature of things; rather, their deliverances need to be checked against and interpreted in the light of our clear and distinct ideas. We can separate dreams from waking life by observing the consistency and coherence of what we really experience, reassured by the thought that a benevolent God would not have left us trapped in a lifelong dream. Knowledge is possible when we coordinate our mental powers carefully, subjecting our confused sensations to the discipline of our innate rationality.

Descartes himself was extremely optimistic about his rationalist system, claiming in his preface that his proofs were 'of such a kind that I reckon they leave no room for the possibility that the human mind will ever discover better ones'. His enthusiasm was not universally shared. Few of his critics doubted his claims about the certainty of one's own existence, but there was considerable resistance to his arguments beyond that point. The theologian Antoine Arnauld immediately worried that Descartes was reasoning in a circle in the Third Meditation, needing to rely on clear and distinct ideas in the course of his proof that clear and distinct ideas were in fact trustworthy (this problem, known as the 'Cartesian Circle', continues to haunt Descartes's reputation among philosophers to this day). Princess Elizabeth of Bohemia pointed out that Descartes was disappointingly far from having a clear account of how the intellectual soul could interact with the physical body. The English philosopher Thomas Hobbes pressed Descartes on the question of how exactly 'innate ideas' were supposed to work. This last criticism was then advanced fiercely by the English philosopher John Locke (1632–1704).

The Empiricism of John Locke

Although Locke (see Figure 5) studied some traditional philosophy as part of his undergraduate arts degree at Oxford, it was only when he came across the new philosophy of Descartes, several years after graduation, that he became seriously interested in the subject. Amid a variety of other pursuits, including the study of medicine and a successful career as a civil servant, Locke worked on a theory of knowledge to rival Descartes's, spending 20 years developing his main work, the *Essay Concerning Human Understanding*. The central goal of the *Essay*, as Locke explains in his introduction to the work, is to 'search out the bounds between opinion and knowledge'. Locke was convinced that he could figure out the limits of what is humanly knowable by using what he called the 'historical, plain method' to study the natural operations of the human mind. Beyond the limits of what could be known, Locke argued, human beings could certainly have faith or opinion, but they should neither claim certainty, nor attack those who differ in matters of opinion or faith (the promotion of tolerance was a large part of Locke's agenda in figuring out the limits of knowledge).

The 'historical, plain method' led Locke to challenge the rationalist admiration of innate ideas. Looking at how human beings actually make judgements, is it really obvious that we have rational ideas installed in us from the start? Rationalists claimed that some principles are embraced by all of humanity and must be innate ('whatsoever is, is'; ''tis impossible for the same thing to be, and not to be'), but Locke's observations of humanity led him to disagree. 'Who perceives not, that a child certainly knows, that a stranger is not its mother; that its sucking-bottle is not the rod, long before he knows, that *'tis impossible for the same thing to be, and not to be*?' Against the suggestion that such principles might be innate, but not yet noticed, at least by children, Locke declared it to be 'near a contradiction, to say, that there are truths imprinted on the soul, which it perceives or understands not'.

5. John Locke, 1632–1704

What could 'imprinting on the soul' even mean, unless it meant making the soul aware of something? Advocates of innateness might try to say that innate truths won't be recognized until a child gains the mature use of reason, but Locke objects that requiring the use of reason to discover a truth makes it redundant to claim that this truth is also innate or built in. It would make as much sense as saying that the idea of the sun is innate within us but also requires the development of vision before it can be noticed by the mind.

Where the rationalists had argued that the mind comes pre-stocked with innate ideas, Locke maintained that our first ideas come from sensation: the mind is 'white paper' before sensation begins to mark it. Through repeated experiences, children gradually become able to recognize objects and people, as their raw 'ideas of sensation' cluster into patterns. As our mental powers develop, we also gain what Locke calls 'ideas of reflection' from witnessing the mind's own operations. These two sources, sensation of outer things and reflection on mental activities, are according to Locke the only sources from which human beings ever gain the simple ideas that form the building blocks of all human thought. This experience-centred approach to knowledge is known as empiricism.

Although empiricist accounts of knowledge insist that the raw materials of thought must come from experience, they leave room for more processed materials from elsewhere. Having received simple impressions from sensation and reflection, we can combine them into new arrangements, forming complex ideas of more elaborate and abstract things like justice, property, and government, or dreaming up imaginary creatures or new inventions. Once we have crafted these elaborate ideas ourselves out of simpler components, we can have clear knowledge of various truths about them. For example, Locke thought that the claim 'Where there is no property, there is no injustice' was 'a proposition as certain as any demonstration in Euclid'. If you can't immediately see why, this could be a sign that the ideas that come to mind when you read the words 'property' and 'injustice' are not quite the same as Locke's. Locke himself defines property as 'a right to any thing' and injustice as 'the invasion or violation of that right'. His 'no injustice without property' line falls out pretty easily if you start with these ways of defining the key terms.

Of course, Locke recognizes that we aren't in fact all defining our key terms the same way, and maintains that the most frustrating debates about moral and political matters arise from people associating different ideas with their words, and not really from any deeper difference in the ideas themselves. If we could gain greater clarity on precisely which ideas we mean by our words, for example by analysing them into simpler terms, then we could clear up these merely verbal disputes and achieve greater agreement and more extensive knowledge.

Knowledge has its limits, however, even when we take the greatest care to define our terms precisely. Locke defines knowledge itself as 'the perception of the connection and agreement, or disagreement and repugnancy, of any of our ideas'. There are various ways in which ideas can be seen to agree or disagree: Locke calls these the 'degrees' of knowledge.

The clearest degree of knowledge is *intuitive knowledge*, and this is what we have when we immediately grasp the agreement or disagreement of some ideas ('that a circle is not a triangle'). A slightly more difficult kind of knowledge is *demonstrative knowledge*, in which the mind sees an agreement or disagreement among its ideas but only with the help of some chain of connecting ideas: our knowledge that the internal angles of a triangle add up to two right angles, for example, is demonstrative knowledge, because it runs through a series of stages.

The last grade of knowledge is what Locke calls *sensitive knowledge*. Sensitive knowledge differs from the other kinds in being concerned not with general truths or relations among ideas, but with the existence of particular objects that we experience. According to Locke, you have sensitive knowledge of the existence of the things you are sensing: for example, the clothes that you are now wearing. You experience something very different, Locke observes, when you are really feeling something as opposed to merely remembering it or imagining it: it's impossible to miss the difference between looking at the sun during the day and just thinking about it at night. Because sensitive knowledge links our ideas to reality, it might seem to violate Locke's general definition of all knowledge as the perception of agreement or disagreement among ideas. But it is possible to see even sensitive knowledge as generated by noticing the agreement of ideas, as long as we remember that ideas include ideas of reflection, generated by our awareness of our mind's own operations. Whenever you have sensitive knowledge (for example, knowledge of the existence of the clothes you are now wearing) you experience your own mental operations of sensing the clothes, which are quite different from the mental operations you'd have if you were just remembering or imagining these clothes. You recognize that this feeling agrees with your idea of sensation, your idea of being in touch with things that really exist in front of you now, rather than your idea of an operation like memory or imagination.

Locke doesn't invest much energy into trying to come up with a clear answer to the sceptic who worries that you might feel you are experiencing something even when you aren't, perhaps in part because he seems to think it's humanly impossible to maintain scepticism sincerely. Locke thinks that we are 'invincibly conscious' of the reality of the things we sense, and can at most pretend to be sceptical about it. Locke also puts a heavy emphasis on the adequacy of our knowledge for our practical purposes: if what seems to be real is real enough to be a reliable source of pleasure and pain, then you can be as certain that it exists as you need to be. For Locke, knowledge is above all a tool for the pursuit of happiness.

Knowledge is not our only guide. Locke argued that many of our actions are governed not by knowledge but by something weaker, namely judgement. Judgement doesn't give us certainty, according to Locke, but it allows us to hold that a claim is probably true—and in many cases, the probability is high enough that we can treat it as practically certain. What we believe on the testimony of others, for example, is according to Locke always a matter of judgement rather than knowledge (a controversial claim, as we'll see in Chapter 6).

Locke's empiricism and Descartes's rationalism emphasize different aspects of the new way of thinking about nature that emerges in the Early Modern period. Descartes focuses on the importance of mathematical and abstract ideas; Locke focuses on the importance of experience and observation. Given that these thinkers both share the Early Modern agenda of supporting the new scientific way of thinking, how did they come to produce such different theories of knowledge? One possible reason is that they attack the problem of knowledge from different angles. In the *Meditations*, Descartes takes very much a first-person approach: his guiding question is: 'What can I know for certain?' Starting from his inner consciousness of his own existence, he then gradually moves outward by means of an inventory of the contents and powers of his own mind. From the first-person angle, the clearest instances of knowledge are those that do not depend on anything in the external environment: we can be confident of purely rational operations on abstract ideas even when we are unsure of the outer world. Locke shares some of Descartes's introspective tendencies, but he is also happy to adopt a third-person approach in many passages, drawing on his observations of others alongside himself. The main question Locke aims to answer is: 'What do human beings know?' How does the child come to recognize the difference between its sucking-bottle and the rod? From the third-person angle, we naturally take for granted the fact that the child is experiencing something real when we see him react to the sucking-bottle. The third-person point of view makes sensory perception, seen as a relation between persons and their environment, a very natural foundation for knowledge.

The ideal, of course, would be to find a theory of knowledge that explains both the abstract and the observational (and how they fit together). Making progress towards that goal seems to require a deeper understanding of the relationship between the first- and third-person points of view on knowledge, a problem that remains a very active research topic in contemporary epistemology, and is the subject matter of Chapter 5. In modern terminology, the choice between taking a first-person or a third-person approach is the choice between 'internalism' and 'externalism'. The importance of this choice became very clear in the 1960s, as philosophers struggled to answer a surprising challenge to traditional ways of analysing the concept of knowledge.

Chapter 4
The analysis of knowledge

Gettier's challenge

Walking through the deserted train station on a Sunday afternoon, Smith realizes he has lost track of time. He glances up at the familiar station clock and sees the hands clearly pointing to 1:17, which is in fact the current time (Figure 6). If you are tempted to say Smith now knows that it is 1:17, wait a moment. Here's one additional fact about the situation: the clock is broken, and its hands haven't moved for the last two days. It's just a coincidence that Smith is looking at this broken clock at one of the rare moments when it isn't wrong. By looking at the broken clock, does Smith come to *know* that it is 1:17? Many people feel inclined to say *No, he doesn't know this.* (If you're not one of them, try the stories in Box 4 to see if they work better for you).

In a book published in 1948, Bertrand Russell offered the broken clock story as an example of a true belief which didn't count as knowledge. Russell included it alongside several other examples, such as a story of an man who buys a lottery ticket, feverishly convinced he will win, and actually does win. It didn't attract special attention at the time, but there is something special about the clock case. Fifteen years later, philosophers went back and took a second look at stories like the one about the clock, noticing a very interesting feature. Unlike the irrationally optimistic lottery

6. A stopped clock

ticket buyer, Smith is being completely reasonable: he has what might seem like perfectly respectable evidence for his true belief about the time. Ordinarily, if someone asks whether you know the time, you just glance at your wristwatch and answer. You don't say: 'I can't be sure whether I *know* what time it is until I double-check that my watch is running properly.'

What prompted philosophers to think twice about the clock story was a very short paper published in 1963 by an American philosopher, Edmund Gettier. Gettier's aim was to challenge the way knowledge was analysed. The leading theories of knowledge in his day all equated knowledge with justified true belief (this equation was called 'the classical analysis of knowledge'). There were, of course, some debates over how the key terms should be understood; for example, over whether justification required a special kind of evidence, or an ability to defend a claim if challenged. But despite these controversies over the details, the fundamental structure of the classical analysis rested unchallenged. Gettier's 1963 article presented two stories which, like the clock story, illustrated the possibility of lacking knowledge even while making a judgement that is both true and justified (or reasonable, or supported by evidence). Philosophers had recognized for a long time that not every true belief counts as knowledge (even Plato talks about lucky guesses); what was new was the observation that even justified true beliefs could fall short.

If the man who looks at the broken clock has justified true belief without knowledge, the classical analysis is really missing something. It's not enough to add some justification to true belief if the justification and the truth of the belief aren't properly

Box 2 The classical analysis of knowledge

(where S stands for a subject or person, and *p* stands for a proposition)

S knows that *p* if and only if

(i) *p* is true;
(ii) S believes that *p*;
(iii) S is justified in believing that *p*.

related to each other: knowledge is something more than the simple sum of justification and true belief.

At first it seemed that the problem with the classical analysis would be easy to fix. One feature of the clock case you may have noticed is that Smith seems to have a false belief: he presumably thinks the clock is working, and he's wrong about that. Can we save the basic idea of the classical analysis by adding an extra condition ruling out reliance on false beliefs on the way to the truth? Perhaps achieving knowledge is like crossing a wooden bridge, where stepping on one rotten plank on the way across will make you fall short of the goal. A rule barring reliance on false belief would ensure that every step is sound.

This no-false-belief proposal was advanced within a few months of the publication of Gettier's article, and then immediately shot down from two different sides. On the one hand, people apparently *can* have knowledge despite the presence of a supporting false belief. Imagine a detective investigating a broad-daylight assault. She interviews a dozen witnesses who all say that they saw Jones hitting Smith, and then she gathers plenty of physical evidence, including blood droplets of Smith's type on Jones's knuckles, gets a statement from Smith, and even a confession from Jones himself. Does the detective now know that Jones hit Smith? Of course she does. (Only a sceptic would say otherwise.) But now try supposing that one of the dozen witnesses was lying in saying that he saw the assault (the others were all there and are being honest). If the detective believes that all the witnesses *including the liar* saw the assault, then she has a false belief supporting her judgement that Jones hit Smith. But it doesn't seem that this false belief is enough to bar the detective from knowing that the assault happened, given the extent of her otherwise solid evidence. The detective's one false belief is like a single rotten plank on a bridge that is so heavily reinforced with strong timber that it is still safe to cross. We don't (and shouldn't) follow a simple rule of denying knowledge every time a false belief enters the picture.

Further tricks can be applied to try to rescue the classical analysis. For example, we could try to say that our detective wasn't essentially relying on the testimony of that one witness: she would have thought the same thing even without him. Perhaps we can save the classical analysis by insisting that the knower can't *essentially rely* on any false belief (it can't be a crucial weight-bearing plank in the bridge). But it turns out to be very hard to explain what it means to 'essentially rely' on something in the course of forming a judgement—a blizzard of rival theories tried to work out that question, and none seemed to succeed. Meanwhile, there was fading enthusiasm for the whole barring-false-belief way of enhancing the classical analysis, because this approach was also failing on another frontier. Although Gettier's own stories happened to involve people with false beliefs, it turns out to be possible—as we'll see shortly—to construct Gettier-type stories in which a character has justified true belief without knowledge despite having no relevant false beliefs at all. While defenders of the classical analysis struggled to find a way to patch it up, others were inspired to propose entirely fresh accounts of the nature of knowledge.

The causal theory of knowledge

Walking into your neighbourhood bakery at the end of the day, craving a muffin, you are happy to see what looks like a full basket of muffins in the display case. 'Thank goodness they haven't sold out!' you think to yourself. What you don't know is that the basket you are looking at is strictly decorative: the 'muffins' you see are plastic fakes, attractively arranged to make the display case look good, and left there day and night, whether or not the store still has muffins available. As luck would have it, there are in fact a few muffins left in the bakery, out of view, on a shelf under the counter. Your belief that there are still muffins in the store is true, and—assuming that the plastic fakes are very realistic—justified. But do you know, on the basis of seeing the display case, that the bakery has not sold out of muffins for the day?

In this Gettier case there is some gap between the source of your justification (the plastic display) and the fact that makes your belief true (the muffins under the counter). A good theory of knowledge would do something to close that gap. In 1967, Alvin Goldman proposed a new and very simple theory promising to do just that. According to Goldman's *causal theory of knowledge*, experience-based knowledge requires the knower to be appropriately causally connected to a fact. What is wrong with your true judgement about the muffins, according to this theory, is that it is caused by your seeing the fakes. The fact that there are still muffins in the store is not what is causing you to believe that there are still muffins in the store. In ordinary cases of experience-based knowledge, on the other hand, the fact we believe has a clear causal connection to our believing it: when you have a clear view of the burning barn, that burning barn is causing you to have visual experiences that cause you to form your belief that the barn is burning. Memory and testimony can also connect us causally with facts, through causal chains linking us back to earlier events in our experience and the experience of others.

The classical analysis insisted that our beliefs have to be justified: this condition is conspicuously absent from the causal theory. Goldman's view is that there are many cases in which people should count as knowing things even if they can't now justify themselves by pointing to any supporting evidence. Average educated adults might know, for example, that Julius Caesar was

Box 3 The causal theory of knowledge

S knows that *p* if and only if the fact *p* is causally connected in an appropriate way with S's believing *p*.

('Appropriate' knowledge-producing causal processes include memory, perception, and inference.)

assassinated, even if they can't remember where they learned this fact, or give you any supporting reasons for their claim. The causal theory can credit them with knowledge as long as their belief that Caesar was assassinated has the right kind of causal grounding in the fact that Caesar was assassinated. An appropriate causal grounding could run from the fact of Caesar's being killed to the perceptual experiences of eyewitnesses to the event, and then through their testimony to ancient historians, carried forward through centuries of history books to the forgotten teachers who caused today's adults to have this belief about Caesar. It doesn't matter whether these adults remember exactly what their sources were; what matters is just that their beliefs actually have an appropriate causal chain anchoring them to the fact believed.

The causal theory was celebrated as a great advance in the theory of knowledge, but it harboured a few problems. Some critics worried about how exactly we could spell out the notion of an 'appropriate' causal chain without using the idea of knowledge itself, the very idea we were trying to define. Others worried that there was something wrong with divorcing knowledge from justification and evidence (more on those particular worries in Chapter 5). But one of the deepest worries came from Alvin Goldman himself, a decade after he published the causal theory. It turns out to be possible to tell a story in which someone *is* causally related to a fact in an appropriate way, but still lacks knowledge.

Here's the story: Henry is driving in the countryside with his young son, identifying landmarks for him as they drive past. 'Look, son—a cow! Over there—a tractor! There's a barn over there, in that field!' Henry's belief that there's a barn in the field is caused by his perceptual experience of the actual barn. By the standards of the causal theory, Henry knows that there is a barn in the field, and (so far) it seems that the causal theory fits common sense on this point, too: we'd ordinarily say Henry knows that there is a barn there, even if he doesn't get out of the car to walk up close to the structure and double-check. Barns are big enough, and distinctive enough, that you can generally tell that something is a barn from quite a distance away.

Once again, however, there's a twist in the case. He doesn't know this, but Henry is driving through Fake Barn County, where the zany locals have put up dozens of barn façades, false fronts that just look like barns when seen from the highway. It's sheer luck that Henry is right now looking at the one real barn in the region: if he had been looking at any of the other similar-looking things in the area he would have formed a false belief in taking it to be a barn. Does Henry know that what he sees is a barn? Goldman thinks not; Henry's risk of error is too high. Notice here that Henry isn't actually forming any false beliefs as he makes his judgement about that one real barn, and (assuming that he has no idea he's in Fake Barn County) he's also justified in believing that the roadside object is a barn. There's nothing wrong with the causal history of Henry's belief, either, so if he doesn't have knowledge, there's something wrong with the causal theory of knowledge.

Goldman at this point launched a fresh analysis of knowledge, focused not on causation but on reliability: the problem with Henry is that his ordinary barn-spotting mechanisms aren't reliable in Fake Barn County. On the new analysis, knowledge is true belief produced by a reliable belief-producing mechanism, where by 'reliable' we mean 'likely to produce a true belief'. Known as reliabilism, this analysis of knowledge rapidly won many fans.

Reliabilism also sparked criticism. One of the main criticisms concerns the key word 'likely'. If knowledge must be formed by a mechanism *likely* to produce true belief, what rate of success should count as likely enough? Would 85 per cent be enough? Or 99 per cent? The problem is not just that there are borderline cases where it would be hard to tell whether or not someone had knowledge. Lots of concepts have tough-to-call borderline cases, so it wouldn't be very worrisome if it were clear that 99 per cent reliability was enough for knowledge, and 97 per cent not enough, and we just had trouble figuring out what to say about 98 per cent. The problem is rather that some factors make it look like we have to raise the line very high, even while other factors drive us to lower it.

To see why a high threshold looks good, imagine that you have a ticket in a fair one-in-a-thousand lottery for a huge cash prize. The draw has just been held, but the winner has not yet been announced. As you wait for the announcement, do you know that your ticket has not won? Odds are 99.9 per cent that it has lost, but most people say that despite the slim odds of winning, they don't know that they have lost until the announcement is made. But if a 99.9 per cent chance isn't high enough for knowledge, how high do our chances need to be? (You can even push the number up from 99.9 per cent just by imagining a larger lottery—even with a million tickets sold, it still seems you don't *know* that your ticket has lost just on the basis of recognizing the long odds against it.) So, does your belief-forming mechanism have to be more than 99.9999 per cent reliable to gain knowledge by using it? If we insist on that, it looks like we are headed down a slippery slope towards scepticism: we want to allow that people can know things when they form true beliefs through ordinary sensory perception, even if it has a more than one-in-a-million chance of delivering a false belief. The reliabilist analysis of knowledge (true belief that is produced by a mechanism likely to produce true belief) is attractively simple, but it's not immediately obvious how to make it work (we'll explore the strengths and weaknesses of reliabilism in more detail in Chapter 5).

No way out?

Gettier showed that the classical analysis was missing something. Philosophers hoped at first that this missing ingredient would be easy to find, that we could just add a fourth condition to the classical three, or devise a new way of breaking knowledge down into simpler building blocks. These hopes have been dashed: in the decades after Gettier, dozens of analyses of knowledge have been formulated, but none has secured broad support. Many analyses turned out to be vulnerable to new intuitive counter-examples; others turned out to be disappointingly circular, essentially assuming the concept they were trying to analyse, sometimes under the disguise of newly introduced technical vocabulary. After three decades of increasingly complicated and unsatisfying proposals, some philosophers began to suspect that the problem of analysing knowledge was not solvable. But why not?

The answer may lie in the relationship between justification and truth. It's widely held that we can be justified in believing falsehoods: unlike knowledge, justification doesn't ensure truth. The person who is taken in by a very realistic holographic projection of a muffin isn't crazy to think that there's a muffin in front of her: she is being perfectly reasonable, even if what she believes is false or only coincidentally true. We could say the same for the investigator who ends up believing something false because he encounters tons of (misleading) evidence: given what he has found, he could be entirely reasonable to believe that the butler is guilty, for example. In fact, the possibility of justified false belief was part of what gave the classical analysis its original punch: if we ask which justified beliefs count as knowledge, the truth condition narrows it down to 'just the true ones'. But once we allow that justification doesn't always lead to the truth, the Gettier problem can start to look inescapable. There's now even a standard recipe for cooking up counter-examples to analyses of knowledge, courtesy of the American philosopher Linda Zagzebski. Here's how it works: describe a situation in which someone has a false but justified belief (or more generally, a false belief that fits all the conditions in the proposed analysis other than truth). Then add a lucky twist to the story so that the proposition believed ends up being true. Zagzebski's recipe has generated counter-examples to a massive range of proposed theories of knowledge.

Can knowing be analysed at all?

If efforts to analyse knowledge aren't going very well, it's useful to step back and ask why. Some philosophers have argued that the real problem here is that *knowledge* is not a well-behaved concept: Matt Weiner, for example, thinks that our use of the verb 'know' is guided by a set of handy but inconsistent principles. Trying to get a clear definition of knowledge out of the conflicting ways we intuitively speak of it is like trying to identify the make and model of a car composed of assorted scrap parts. Other philosophers have argued that we should shift our attention away from knowledge and refocus on other topics, like the question of what it is reasonable to believe. But it's not obvious that the difficulties we've experienced in analysing knowledge should lead us to turn away from it. Sometimes resistance to analysis is a good sign, a sign we have reached something fundamental.

To see the value of resistance to analysis, consider a case where resistance is low: table salt can easily be analysed as sodium chloride (NaCl). The building blocks here (Na^+ and Cl^-) are simpler than the compound they form together, and these basic building blocks can be combined with other materials to form different compounds. The project of analysing knowledge started with the idea that knowledge was also composed of simpler materials, including truth and belief. It's clearly right to say that truth and belief can be combined with other materials to make states that differ from knowledge—when beliefs are combined with factors like falsity, they fail to count as knowledge, and so on. But the other key assumption in the analysis-of-knowledge project is more controversial. Are the building blocks of belief and truth really simpler than knowledge itself, and is knowledge really a compound built out of those factors (perhaps with some other factors thrown in)?

Maybe not. In particular, it's not obvious that *believing* is simpler and more basic than *knowing*. What if *knowing* is the fundamental idea, and *believing* is a spin-off from it? This idea is championed by the 'knowledge-first' movement in epistemology. Leading this movement, Timothy Williamson argues that one reason why philosophers have not been able to come up with a satisfactory analysis of knowledge in terms of belief plus further factors is that the concept of knowledge is more fundamental or basic than the concept of belief.

There's something initially counter-intuitive about the knowledge-first approach. It can seem that believing just has to be a more basic building block for the simple reason that states of belief are more plentiful than states of knowledge. Every time someone knows something, they can also be described as believing it, but not vice versa: beliefs that are false or irrational, for example, aren't knowledge. To go back to our original analogy, basic building blocks are in many cases more common than the compounds they form: sodium, for example, is more common than table salt because sodium is present not only in sodium chloride but also in sodium nitrate and many other compounds (whenever you have table salt you have sodium, but not vice versa). However, it's not always true that the more common thing is more basic. Consider the concept of a perfect circle: very simple, and perhaps quite rarely found in nature. Now, if by the word 'rounded' we mean 'at least roughly circular', then there will be many more rounded things than circular things, but the concept of the circle would still be more basic. The circle is our fundamental starting point, something we used in defining 'rounded', and for good reason: the clean geometrical nature of circularity is simpler than the messy geometrical nature of approximations of circularity.

Box 4 Ancient Gettier cases

In Western philosophy, 1963 is taken as the date of the discovery of cases illustrating the gap between justified true belief and knowledge. But in a text that dates to around the year 770 CE, the Indian philosopher Dharmottara offers the following cases:

A fire has just been lit to roast some meat. The fire hasn't started sending up any smoke, but the smell of the meat has attracted a cloud of insects. From a distance, an observer sees the dark swarm above the horizon and mistakes it for smoke. 'There's a fire burning at that spot,' the distant observer says.

➢ Does the observer *know* that there is a fire burning in the distance?

A desert traveller is searching for water. He sees, in the valley ahead, a shimmering blue expanse. Unfortunately, it's a mirage. But fortunately, when he reaches the spot where there appeared to be water, there actually is water, hidden under a rock.

➢ Did the traveller *know*, as he stood on the hilltop hallucinating, that there was water ahead?

These cases involve a belief that is true and based on what could seem to be good evidence, and just like Gettier, Dharmottara uses these cases as counter-examples to rival theories of knowledge. Many cases like these were actively debated by Indo-Tibetan epistemologists following Dharmottara. Some of the proposals that emerged in Western philosophy since 1963 to handle these cases appeared in the Indo-Tibetan tradition centuries earlier. For example, a detailed causal theory of knowledge was advanced by Gaṅgeśa in the 14th century.

The relationship between knowing and believing is in many ways similar to the relationship between the circle and the at least roughly circular: knowing is the ideal, and believing is some kind of approximation to that ideal. According to the knowledge-first programme in epistemology, it is a bad idea to try to analyse knowledge in terms of belief plus further factors, just as it would be a bad idea to try to analyse the concept of a circle in terms of roundedness plus further factors. According to Williamson, rather than trying to explain knowledge as a compound state formed by adding various factors to belief, we should explain believing in terms of knowing: 'Believing *p* is, roughly, treating *p* as if one knew *p*.' In his view, knowing is a state of mind that essentially involves being right; believing is a state of mind that ideally aims at being right, while perhaps falling short of that target.

Because so much recent epistemology has been devoted to the task of analysing knowledge in terms of true belief plus further factors, it's somewhat revolutionary to declare knowledge to be more basic than belief. If analysing knowledge were the only thing that an epistemologist could do, then declaring knowledge to be basic (or resistant to analysis) would be a way of stopping epistemology in its tracks. However, even if knowledge is not a compound of belief and other factors, there are many things to be learned about it: for example, we can study how it relates to justification, how it is generated and transmitted through processes like perception and testimony, and how it appears when viewed from different perspectives. These issues continue to matter for both knowledge-first and belief-first approaches in epistemology. One of the questions that is agreed to be very important to both sides (and even to the debate between them) concerns the difference made by shifting from a first-person to a third-person perspective. The next chapter investigates the significance of this shift.

Chapter 5
Internalism and externalism

The first-person point of view

Here's something you probably took yourself to know before reading it here: Mount Everest is the tallest mountain in the world. But if that fact about Everest didn't come as news to you, here's something you probably don't know: how exactly you originally learned that fact (or any random trivia fact of that sort). According to psychologists who study memory, unless you had a pivotal life experience when you first heard that fact about Everest (like an earthquake hitting at the very moment it was mentioned in your first primary school geography lesson), you won't remember which source you learned it from. In fact, if someone challenged your claim to know that Mount Everest is the tallest mountain in the world, you might not be able to say much to defend it. You could say that it feels to you like a familiar fact. The challenger could object that those feelings of familiarity can be deceptive. For many people around the world, the claim that Sydney is the capital of Australia feels like a well-known fact. Sometimes, feeling sure you are right accompanies actually being wrong.

Suppose that a person isn't conscious of anything really justifying his claim that Mount Everest is the tallest mountain in the world. Could he still count as knowing that fact? Here philosophers split into two camps. The internalist camp says: If you really can't think of any supporting evidence, you are in trouble. Your belief about Everest can't count as knowledge if there is nothing accessible to you that supports it. It's unlike your belief that you are now reading a book, which you yourself can justify by appeal to the experiences that you are now conscious of having; it's also unlike your belief that there is no largest prime number, which you can justify by going through the steps of Euclid's proof for yourself (here's hoping you remember those steps). Knowledge is grounded by your own experience and by your own capacity to reason. Internalists place a special emphasis on what you can do with resources that are available from the first-person perspective: if you can't see for yourself why you should believe something, you don't actually know it. The subject's own awareness of good grounds is an essential part of what distinguishes knowing from lower states like guessing.

Meanwhile, according to the rival externalist camp, knowledge is a relationship between a person and a fact, and this relationship can be in place even when the person doesn't meet the internalist's demands for first-person access to supporting grounds. If it really is a fact that Mount Everest is the tallest mountain in the world, and if you really are related to that fact in the right way (more about 'the right way' shortly), then you know that Mount Everest is the tallest mountain in the world, even if you can't explain your reasons for thinking this. Externalists are happy to grant that *sometimes* you not only know something but also have special first-person insight into exactly how you know it. But from an externalist perspective, that insight into how you know is an optional bonus, and not something that must generally accompany every single instance of knowledge. Externalists argue that *always* demanding insight into how we know risks setting off a vicious regress. On the internalist way of thinking, they note, you shouldn't just have some random idea about how you know something, but should actually *know* how you know it (what is insight without knowledge?). But if knowledge always requires knowing how you know, then this second level of knowledge requires its own internalist guarantee (knowing how you know that you know), and so on. Suddenly you need infinite levels of insight to know the simplest fact. The internalist path threatens to lead us to scepticism, externalists suggest.

What part does the first-person perspective really play in knowledge? Is it indispensable to distinguishing knowledge from mere belief, or does insisting that it is indispensable lead to scepticism? Although there are signs of a struggle between internalist and externalist tendencies in epistemology as far back as the 18th century, direct debate about this question only emerged after Gettier's challenge to the classical analysis of knowledge as justified true belief. This challenge raised some serious worries about the power of justification, and prompted the development of prominent externalist accounts of knowledge like Goldman's early causal theory and his later theory, reliabilism, discussed in Chapter 4. To get a broader sense of how externalism works, it's worth examining one more influential externalist position—Robert Nozick's tracking theory of knowledge.

Nozick's tracking theory

Here's the fundamental idea behind Nozick's externalist theory of knowledge: the person who knows something not only has the right answer to a given question, but also *would* answer that particular question the right way, even if the answer were different. Imagine a doctor who tells you that you have contracted Hepatitis A. If this doctor just says this blindly to every patient she has, getting it wrong most of the time, then she won't count as knowing you have the disease (even if she's right, and you do). To count as knowing, your doctor should be giving a positive diagnosis to infected patients but not to the ones who are well. What Nozick's theory says is that there is nothing more to knowledge than this: if you have the tendency to believe something when it's true, and not to believe it when it's false,

Box 5 The tracking theory of knowledge

S knows that *p* if and only if:

(1) *p* is true;
(2) S believes that *p*;
(3) If *p* were not true, S would not believe that *p*;
(4) If *p* were true, S would believe that *p*.

then you know it. The theory imposes no special condition about awareness of one's reasons or grounds. The doctor who just has very reliable diagnostic instincts about Hepatitis A, even if she can't explain exactly what features of a patient she is responding to, will count as knowing whether you have the disease when she correctly diagnoses you. She can even count as knowing if she has a *mistaken* idea about how she's making up her mind. Suppose she thinks she's relying on the lab reports when really her judgements are based on subtle cues in the skin tone and smell of her patients: the tracking theory says that as long as her diagnoses actually track whether infected patients have the disease, then she still counts as knowing that they do.

The basic structure of the tracking theory is set out in four conditions: Conditions (1) and (2) are about what is actually happening. Conditions (3) and (4) are subjunctive conditionals covering what *would* happen even in circumstances somewhat different from the actual ones. According to condition (3), your belief that, say, you are now reading a book should be formed in a way that would be generally sensitive to situations in which you are not reading: it has to be the case that you *wouldn't believe you were reading* if in fact you weren't. The person who has constant delusions of reading doesn't know that he's reading even when he is. Meanwhile, according to condition (4), your belief that you are now reading has to be formed in a way that makes you generally alert to the positive activity of reading. It can't be some one-off fluke that you are right about what you are doing this time; you have to be generally a reliable witness about whether you are reading in order to know right now, in this case, that you are. Conditions (3) and (4) are the 'tracking' conditions on knowledge: in order for a true belief to count as knowledge it must be formed in a way that tracks the facts. Your knowledgeable beliefs vary in step with how things are in the world around you; not only do they match the world right now, they are formed in a way that matches how the world would be if things happened slightly differently. If you know, you are securely latched on to the truth.

The tracking theory has some tricky problems of its own. One problem concerns the methods we employ to gain knowledge. Nozick himself realized that there are situations in which the truth of a belief is awkwardly entangled with the method that we use to form that belief. His example runs as follows. Imagine Granny is in hospital, and her darling grandson comes to visit her. As he stands at her bedside, Granny looks at him and forms the true belief that he is well (or at least well enough to visit her). However, we can imagine that the family is so concerned about Granny that they would not let her find out if the grandson were not well (in that case, they'd tell her he was fine, so as not to worry her). So, if the proposition Granny believes (that her grandson is well) were actually false, Granny would be none the wiser: she'd go on believing he was well, violating condition (3) of the tracking theory of knowledge. Somehow it doesn't seem that this secret family plan about what to do if the grandson is ill should stop Granny from knowing that he is actually well enough to visit, as long as he is actually well and standing right there by her bedside. To deal with this problem, Nozick proposed an amendment of the tracking theory to include reference to a 'method of belief formation'. What the Granny case shows is not that there is any problem with Granny's method of belief formation (looking at her grandson at her bedside); the circumstances in which the grandson is ill are not circumstances in which her method would fail to deliver the truth, but circumstances in which she would no longer be in a position to use this method (she'd be using another method, trusting the testimony of her caring but deceptive family).

Like Goldman's reliabilism, Nozick's tracking theory ends up leaning heavily on the notion of a method of belief formation: you know when you hit the truth through a method of belief formation that measures up to certain standards. According to externalist theories, you don't have to know what those standards are; you don't even need to know what method of belief formation you are using. But it does have to be a fact that your method or mechanism of belief formation, whatever it is, actually tracks the facts (on Nozick's tracking theory) or is reliable in the sense of delivering a high proportion of true beliefs (on Goldman's reliabilism). But how is it decided what method a person is actually using in forming a belief? Externalism faces a difficulty here, acknowledged by Goldman as he first formulated reliabilism, and emphasized by internalist critics ever since. This difficulty has become known as 'the Generality Problem'.

The Generality Problem

Remember Henry, driving through Fake Barn County in Chapter 4? Goldman argued that although Henry's belief that what he sees is a barn is actually caused by the right thing (what he sees really is a barn), Henry doesn't know that what he sees is a barn. He's just lucky to be looking at the one real barn in a county full of fakes. Goldman's diagnosis was that Henry's barn-shaped-object-discrimination mechanism is not reliable in this context. In Fake Barn County, that mechanism delivers a false belief almost all of the time, so even though it's delivering a truth for Henry right now, he doesn't count as knowing: knowledge requires a reliable mechanism. It all works out neatly if we follow Goldman in saying that the context that counts is Fake Barn County (and not, say, the world as a whole), and that the mechanism that Henry is using is specifically his barn-discrimination mechanism. Reliabilism would not be able to deliver the desired intuitive verdict if we were to say instead that Henry is relying on a much more general mechanism, like 'vision', which could still be generally reliable, even in Fake Barn County. (Henry uses vision to tell that he is looking at the dotted line on the road, and the trees, and so on, and he's right about all of that—the County doesn't undermine the reliability of his vision as a whole.) We need to be more specific about the mechanism and context to get the result that Henry's belief formation is unreliable. But if we are being more specific, why not go all the way and identify the mechanism as 'barn-shaped-object discrimination, as applied to the barn-shaped object in the very spot Henry is now looking at'? Given that Henry is looking at a real barn, that maximally specific mechanism always gives Henry a true belief. But if we carve up belief-forming processes so narrowly, then any true belief will count as knowledge. How do we hit the target of describing the mechanism and its context at just the right level of detail?

Externalists have proposed many possible ways of answering the Generality Problem. One way already mentioned (if not endorsed) by Goldman in 1976 is to concede that there might be no simple objective fact of the matter about how to describe the mechanism Henry is relying on, because there is no objective fact of the matter about whether Henry knows. Looking at him the way Goldman does at first, Henry doesn't know, but if we take a broader view, perhaps he does. Theories which allow diverse verdicts are complex enough that we will address them separately, in Chapter 7. Another route externalists have pursued is to argue that there really is a natural way of drawing the lines around the process responsible for belief formation in any given instance. Many different proposals have been advanced, drawing on everything from patterns in natural language to the science of belief formation. Externalists consider this a vital and thriving area of research; internalists are sceptical that any satisfactory answers will be found.

Meanwhile, other externalists have argued that the Generality Problem is not exclusively a problem for externalists: internalists may also have to face it at some point. Internalist theories don't focus conspicuously on methods of belief formation: they talk much more about the evidence available to a person to justify or support what that person believes. But even internalists need to care about the manner in which a belief is actually formed, because all epistemologists care about the difference between having good reasons to believe something and actually believing it on the basis of those good reasons. Imagine a juror who has plenty of evidence that a given defendant is guilty of a crime, and believes that the defendant is guilty, but does so on the basis of racial prejudice rather than evidence. Epistemologists of all stripes would want to be able to criticize this juror's belief formation as ill-grounded. But as soon as we care about the processes actually responsible for producing your belief—and not just the evidence accessible to you—we will have to start developing some account of those processes, and we will have to find some way of describing them that is neither too precise nor too general. The Generality Problem seems to be everyone's problem, even if it is more conspicuously a problem for externalists.

Questionable methods

Internalists differ from externalists in assigning a special place to the subject's own point of view. To count as knowing, internalists say, the subject has to be able to see something for herself, simply by reflection or some form of immediate awareness. Different kinds of internalism differ on what it is that the subject has to see here, whether it is some justifying evidence, or completely justifying evidence, or whether the subject has to be able to know that she is justified. But the core idea is that the subject has to be thinking rationally, and making up her mind on the basis of evidence that is accessible to her. Externalists allow for the possibility of knowledge without evidence accessible to the subject, and they argue that this move enables them to defend the status of many beliefs we'd ordinarily want to classify as knowledge (like your belief about Everest being the tallest mountain). Internalists can go either way on that particular example. Some argue that it shouldn't count as knowledge (for most of us, anyway—perhaps experts in mountain geography truly know it). Internalists can say that we are speaking loosely when we describe ordinary adults as *knowing* that Everest is the tallest mountain, just as you might, speaking loosely, describe the shape of France as hexagonal—it's close enough for everyday purposes even if it's not literally a hexagon. Other internalists contend that we do in fact have internally accessible justifiers for beliefs of that sort: one can be aware that one's memory is good, that one has heard this claim many times, or read it in many sources, and so on. Even if some commonly remembered claims (for example about the capital of Australia) are wrong, internalists just need to say that the knowing subject has access to some justifying evidence; most internalists stop short of saying that the subject needs to have access to evidence of a kind that absolutely guarantees the truth of what is believed.

Internalism doesn't necessarily require the subject to be in the special insulated position of being able to know that he knows, or even to have access to evidence that guarantees the truth of what is known: milder forms of internalism can simply insist that the subject should have access to supporting evidence. Here's a story—from American epistemologist Laurence BonJour—to illustrate the desirability of requiring evidence for knowledge. Imagine someone ('Samantha') who thinks she has clairvoyant powers with respect to some odd topic. Samantha thinks that she has a supernatural ability to 'see' where the president of the United States is located at any given moment, although she has no good reason to think that she has this power. She hasn't ever checked up on the accuracy of her paranormal 'sightings', and doesn't have any good reason to believe that clairvoyance is really possible. One day she comes to believe that the president is in New York City, despite reading in her morning paper that he is at home in the White House that day, and despite turning on her television and seeing him give what is apparently a live press conference in Washington, DC. She dismisses all that ordinary evidence and goes with her sense of paranormal vision instead. Does Samantha know that the president is in New York?

Before you answer, here are some crucial twists in the story: the president really is in New York, and all the 'ordinary evidence' that the president is in DC is actually concocted to mislead the public in the face of a national security threat. Plus, despite having no good evidence about her special powers, Samantha is actually right: never mind how, she *does* have reliable supernatural access to the location of the president. On standard externalist theories, BonJour observes, Samantha is going to count as knowing: she is right and reliable. If her belief is formed by a reliable mechanism, it's not just a lucky guess. However, Samantha is being unreasonable in dismissing evidence that goes against her judgement, and in clinging to the deliverances of clairvoyance when she has no supporting evidence about its reliability. It's awkward, to say the least, when a theory of knowledge declares that someone who is being very irrational about some fact actually knows that fact.

Externalists have developed a number of different responses to BonJour's challenge. One possibility is to grant that it seems weird to declare Samantha to have knowledge, but then to insist that she really does know. The feeling of weirdness could be coming from the unfamiliarity of paranormal powers, or our learned cultural attitudes towards those who claim to have such powers. Externalists can also point out that any of us is really in Samantha's position when we start to form beliefs based on the deliverances of our ordinary senses: we don't have good reason to think that vision is reliable before we start to trust vision as a source of information. It remains disquieting that Samantha is ignoring contrary evidence, however. Another externalist response highlights that last point in order to challenge BonJour's claim of having developed a story in which someone is thinking reliably without gaining knowledge. If Samantha is deliberately ignoring counter-evidence to her judgement, then she's thinking in a way which doesn't generally lead to the truth (people who deliberately ignore evidence are generally unreliable). On this way of looking at the case, Samantha's belief formation does start out with one reliable component (the input of paranormal vision on the location of the president), but when we add in the unreliable way she is thinking when she ignores evidence, the total mechanism responsible for her belief formation is unreliable. If your overall method doesn't perform well by externalist standards, then the externalist has an explanation of why it seems you lack knowledge. Naturally, this way of answering BonJour's challenge leads us straight back into the thornbush of the Generality Problem (how do we figure out which way of describing the mechanism is the right one?).

A less confrontational response would be to admit that the internalist has a point here, and to soften externalism by adding an extra condition: instead of just requiring reliability or truth-tracking, one might also require that the subject not have any accessible counter-evidence of a kind that she would be irrational to ignore. Internalists have objected to this effort at compromise: why should we be so sensitive to the first-person availability of counter-evidence if we don't also insist on the first-person availability of positive supporting evidence? Another compromise position grants that accessible evidence is important for certain kinds of belief formation, while insisting that this is not because of a general first-person condition on knowledge. Contemporary psychology recognizes a split between two ways of thinking. We have a rapid, automatic way of thinking which delivers an answer without our being conscious of any sequence of steps (what is 7 times 11?—the answer just pops to mind). This is the kind of thinking that enables you to recognize the face of a friend or state a familiar fact. We also have a slower, controlled way which works through a sequence of stages (what is 7 divided by 11?—the answer takes work). This is the kind of thinking that enables you to solve a sudoku puzzle or figure out your income tax.

It's essential to the operation of this second way of thinking that you have first-person access to a series of steps. Many problems can be tackled either way: if you are asked what you ate for dinner yesterday, for example, you could just answer automatically and say the first thing that comes to mind (and you'll probably get it right), or you could reflect on it carefully and try to reconstruct the events around your meal and revive the detailed experience of eating it. We typically use the automatic way of thinking by default, but we can shift up to the more systematic way if we become self-conscious or anticipate needing to defend our responses (which is just the frame of mind encouraged by systematic reflection on knowledge, and the frame of mind we'd expect Samantha to be in when she struggles with contrary evidence). Internalists may be cheered by the observation that first-person accessible evidence is vital to systematic thinking, but externalists could also take comfort in the thought that automatic thinking is also widely taken to be a legitimate source of knowledge (for example, when you recognize a friend). Externalists can also argue that the reason why first-person accessible evidence is important to systematic thought is just that it is essential to the reliable operation of that mode of thinking.

As we move towards having a deeper understanding of what exactly is available to us from the first-person perspective, there is hope that we'll make progress towards figuring out the true importance of that perspective to the acquisition of knowledge. Meanwhile, progress can also be made by examining rival internalist and externalist theories of various special kinds of knowledge. One special kind of knowledge that is particularly useful to examine here is the kind of knowledge that is delivered through testimony.

Chapter 6
Testimony

They told you so

In the realm of knowledge, many of our prized possessions come to us second-hand. We rely on others for our grasp of everything from the geography of distant places to mundane facts about the lives of our friends. If we couldn't use others as sources, we would lose our grip on topics as diverse as ancient history (except what we could discover through our own personal archaeological expeditions) and celebrity weddings (unless we start getting invited). Testimony evidently expands our horizons: the challenge is in explaining exactly how (and how far). Does listening to other people—or reading what they have written—supply us with knowledge in a unique or distinctive way? Do we need special reasons to trust people in order to gain knowledge from them? What should we think about resources like Wikipedia, where most articles have multiple and anonymous authors?

At one extreme, some philosophers have argued that testimony *never* actually provides knowledge (John Locke will be our star example of this position). At the other end of the spectrum, some philosophers argue that testimony not only provides knowledge, but does so in a distinctive way. In this view, testimony is a special channel for receiving knowledge, a channel with the same basic status as sensory perception and reasoning (this type of position was embraced in classical Indian philosophy, and is now popular in Anglo-American theory as well). By exploring both extremes, as well as leading middle-ground views, we can identify the factors that are broadly agreed to matter most to how we absorb what people say.

No way to know

When does testimony supply knowledge? Some philosophers say: 'Never.' To see why philosophers might be sceptical about testimonial knowledge, even if they aren't sceptical about other kinds of knowledge, it first helps to clarify what we mean by 'testimony'. In an act of testimony, someone tells you something—through speech, gestures, or writing—and the content of what they are telling you plays a special role in what you get out of the exchange. Even sceptics about testimonial knowledge can agree that ordinary perceptual knowledge can be generated by the event of hearing or reading what someone says. For example, imagine either seeing that someone has written 'I have neat handwriting' on a slip of paper, or hearing someone saying 'I have a hoarse voice.' If you can indeed see that the writing is neat or hear that the voice is hoarse, you come to know the truth of what is said or written. But your knowledge here is perceptual, rather than testimonial, because the content of what is written or said plays no special role in what you learn: the sentence 'Smith got the job' would work just as well to convey the beauty of the handwriting or the roughness of the voice. If you believe something on the basis of my testimony, you understand what I am saying, and take my word for it.

For John Locke, there was a sharp contrast between perceptual knowledge (for example, the knowledge that a voice you are now hearing is hoarse) and whatever it is we get through testimony (for example, the news that Smith got the job). The key difference is certainty, which for Locke is a necessary condition for knowledge. Because perception can make you immediately certain of something, as certain as you are intuitively that red is not black, you can gain knowledge perceptually. What we get from testimony, Locke says, is at best highly probable, as opposed to certain. During an English winter, if you see a man walking across an icy lake, then you know that this man is crossing the lake. If someone else tells you that she has seen a man walking across the lake, then as long as your informant is trustworthy and what she says fits your own past observations, it is rational to consider the report to be very likely to be true, but you will not actually know that it is.

A key role is played by your own background experiences. Locke tells the story of the King of Siam hearing from a Dutch ambassador that water in Holland becomes solid enough in winter to support the weight of a man, or even an elephant (if you could coax an elephant to Holland in the winter). The king is said to have replied, 'Hitherto I have believed the strange things you have told me, because I look upon you as a sober fair man, but now I am sure you lie.' Locke is sympathetic to the doubting king: given all the king's past experiences in the tropics, it is entirely reasonable of him to find it more probable that the ambassador is lying than that water ever naturally becomes solid.

Even if testimony is never quite certain, it can still be more or less likely to be true, and Locke counsels maintaining a level of confidence in testimony that reflects the strength of your evidence. He has a complex formula for determining the perfectly rational degree of confidence in testimony: after first weighing how well it fits with your own experience, you must take into account the following six factors:

1. The number of witnesses
2. Their integrity
3. Their skill
4. The purpose they have in supplying their report
5. The internal consistency of what is conveyed, and the circumstances of your hearing it
6. Whether there is any contrary testimony

While he thinks testimony doesn't transmit knowledge, Locke doesn't think we should generally resist what others tell us: he says that the reasonable person will assent to testimony. When it fits with our own observations and scores highly on his six-point checklist, what we get from testimony is for practical purposes very like knowledge ('we receive it as easily, and build as firmly upon it, as if it were certain knowledge'). But what we get from testimony is unlike the discovery that red is not black, according to Locke, because it is open to being undermined later by further experiences. (You thought you had no reason to distrust her, but the woman who told you she saw a man crossing the lake was making it up, and later you hear from ten other people that the ice was too thin to cross today.) Locke argues that this vulnerability to future contrary reports means that what we get from testimony doesn't literally count as knowledge.

We'll look at a challenge to Locke's reasoning shortly. But first it's worth taking a moment to appreciate how radical his position really is. If Locke is right, then the proper answer to the question 'Do you know where you were born?' is 'no' (assuming that your beliefs on this matter are, like most people's, determined by what your family has told you, or what is written on your birth certificate). You could say that it is very probable you were born in a certain place, but not having retained first-hand experience of the location, you won't have knowledge of this fact. You also don't know that George Washington was once the president of the United States, or that Antarctica exists (assuming you haven't been there yourself). Lockean readers of this book can't even describe themselves as knowing that John Locke ever lived: they should at most consider it highly probable that he did.

It's clear that Locke is going against the way we ordinarily speak: we very freely describe people as gaining knowledge through testimony. ('Does Jones know that Smith got the job?'—'Yes, he does—I just told him.') However, we can recognize that in many situations the way we ordinarily speak is not strictly accurate (for example, when we talk of the sun rising or setting when really it is the earth that is rotating). Does Locke have good reasons of principle for saying that strictly speaking we do not gain knowledge from testimony? His argument about vulnerability to later doubts is questionable, in part because it seems to apply equally well to judgements grounded in perception and memory, which he does want to classify as knowledge. Locke thinks that perception enables you to know, for example, that you are now reading a book, and also to remember later that you were reading a book, retaining your perceptual knowledge through memory. However, it's possible here too that later on you will come to doubt yourself. Even if you are really perceiving something now, you might have doubts later on, perhaps wondering whether you were only dreaming. Locke doesn't seem to think that the possibility of stirring up later doubts should undermine your claim to know right now, as long as you are actually now perceiving and not dreaming. But a parallel argument could be applied to testimony: if someone knowledgeable tells you that Smith got the job, and you don't actually have any doubts about what they are saying right now, then you should now have the certainty needed for knowledge. If you start doubting later on, for example because of contrary testimony, you could lose that knowledge, but this is not proof that you never had it. If your informant was knowledgeable, then your later doubts couldn't show that what you originally judged was untrue: if your informant knew that Smith got the job, then it must be true that Smith got the job. Any report to the contrary is misleading. Of course, there could be situations in which you fail to have doubts, and take the word of a liar as if she were telling the truth, but these situations are parallel to situations in which you are taken in by a perceptual illusion. If there is a big difference between the knowledge-providing powers of perception and testimony, Locke hasn't shown us what it is.

The middle ground: reductionism

In claiming that testimony never supplies knowledge, Locke occupies a minority position. Most philosophers are more positive about it. The main moderately positive position is reductionism: we do gain knowledge through testimony, but the knowledge-providing power of testimony is nothing special. Whether we are reading, listening, or watching someone's gestures or sign language, we receive testimony through ordinary sense perception. Assuming everything goes well, sensory perception lets us know that a speaker has said a sentence. In order to gain knowledge of what the sentence itself says, and not just the fact that the speaker said it, we rely on our ordinary powers of inference and perception. There is still something Lockean about this: you look at the sort of factors on Locke's checklist for the likelihood of testimony being true (how well it fits with past experience, evidence about the integrity of the speaker, and so on), but when what you are hearing scores high enough, you come to know the communicated proposition. This way of thinking about testimony is known as 'reductionism', because the knowledge-providing power of testimony can be reduced to the knowledge-providing power of other sources, notably perception, memory, and inference.

Reductionism comes in two flavours: global and local. According to global reductionism, your own experience of the world gradually teaches you that testimony, in general, is a fine source of knowledge. As a youth, you ask for directions to the train station, someone tells you, and then even if you don't yet know the truth of what they said, you can follow the directions and confirm their truth for yourself. Because you can often double-check the truth of what people say, over time you gain knowledge of the track record of past testimony, which then works as a positive experience-based reason to accept present testimony. An ordinary adult can know that the train station is down the road and to the right as soon as he is told, not because testimony has any distinctive knowledge-generating power, but because his own past perceptions, memories, and inferences support accepting what he now hears. The global reductionist doesn't have to say that you should believe absolutely everything you hear: if you are in a situation where there are special undermining factors—for example, if you know that the person you are talking to has a strong incentive to lie—then you can take that into account. But when there are no special warning signs, the global reductionist says you have a standing positive reason to believe what you are told.

The local reductionist tries something more finely tailored: rather than seeking a blanket positive reason to trust all testimony, the local reductionist suggests that you look for specific positive reasons, in any given situation, to accept the word of the person you are hearing on the topic she is speaking about. Is this person an expert? Has she told you the truth in the past? How plausible is her story now? Again the specific reasons we rely on ultimately come from perception, inference, and memory, rather than on testimony itself. If these ordinary reasons are strong enough in a given situation, you can know the truth of what you are being told.

Both forms of reductionism allow us to say that most adults know where they were born and know that Antarctica exists. When testimony comes from close and trusted witnesses (like your parents, telling you about your birthplace) or from appropriate experts (numerous map-makers and travel writers, telling you about a distant continent), then it can supply knowledge, according to both forms of reductionism. In cases where you don't have any special reasons to trust your informant—lost in a strange city, you ask a total stranger for directions—the global reductionist can say that you gain knowledge, but the typical local reductionist cannot. (If you are lucky, the local reductionist says, you gain a true belief.)

Local reductionism can sound very calculating: in practice, we don't often weigh the reasons to trust someone before accepting their word. However, local reductionism about testimonial knowledge is not a descriptive theory about how we actually form our beliefs in daily practice: it's a theory about the conditions under which those beliefs deserve to count as knowledge. Even if we tend to trust strangers blindly when we ask them for directions, the local reductionist suggests that we shouldn't think of ourselves as gaining knowledge on this basis. Attaining knowledge requires greater caution, if local reductionism is right. Taking this line requires explaining just why we need greater caution for testimonially grounded knowledge than for knowledge grounded in perception and reason. It's true that testimony can let us down (sometimes our informants are dishonest or confused) but perception can also let us down (sometimes our eyes play tricks on us). One possible reason why testimony could be special is that it involves free agents who have purposes of their own. Human communication differs from the communication among bees, for example, who reliably signal to each other the location of nectar-bearing plants. A bee who learns the location of nectar from another bee is able to fly there as well as if it had witnessed that place: bee signalling gives bees the benefit of each other's experience, in what is sometimes called 'cognition by proxy'. Bee signals can be defective (if the bee is sick, or if the nectar-bearing plants are moved by an interfering researcher after the first bee's contact), but these defects are like the defects in our perceptual organs (when we are sick, or when things are moved behind our backs). One reason that bees can gain the benefit of other bees' experiences directly is that they cannot deliberately deceive. A local reductionist could stress that caution becomes important for communications between members of a sneakier species like ours.

Alternatively, local reductionists could dispute the suggestion that we often rely blindly on the advice of strangers. Perhaps we do ordinarily exercise caution, but in ways that are subtler than explicit weighing of the reasons to trust someone. Recent empirical work on 'epistemic vigilance' has advanced our understanding of how and when we actually accept the word of others. Even if we aren't explicitly thinking to ourselves about the reliability of the stranger we've asked for directions, we could be monitoring his facial expressions and speech patterns to assess how trustworthy he is. Better insight into our actual practices can help us see whether the local reductionist is in fact proposing an account that fits those practices quite closely, or instead proposing that our actual practices are sloppy and we don't have knowledge as often as we think.

Testimony as a distinctive source of knowledge

A still more generous approach to testimony is possible. Instead of seeing testimony as dependent on other ways of knowing, such as past experience and reasoning, you might think of it as a basic source of knowledge in its own right. According to the direct view of testimony (sometimes also called the 'default' view), when your knowledgeable co-worker tells you that Smith got the job, you know that Smith got the job, and your knowledge isn't dependent on your reasoning about the track record of testimony or the reliability of that particular co-worker. You just have to understand what a knowledgeable informant is saying in order to gain knowledge. It's true that the testimonial channel takes input from sensory perception (you have to be able to hear what someone is saying, or read what they have written); reasoning also takes input from sensory perception (for example, you look at a half-solved sudoku puzzle and start to work out the answer). But testimony remains a distinct way of knowing something, just as reasoning is distinct from pure sensory perception. The way you think when you understand what someone says is different from the way you think when you see something with your own eyes, and different again from the way you think when you are engaged in reasoning or puzzle solving.

The direct view of testimony dates back a long way: it is defended by the Indian philosopher Akṣapāda Gautama in the 2nd century CE. Gautama maintains that testimony is a special channel through which we gain knowledge, and emphasizes that testimony is not a form of inference. We do not think to ourselves: 'Lee has said that Smith got the job, and Lee is a reliable person, therefore Smith got the job.' We know, as soon as we understand what Lee has said, that Smith got the job (we can also focus on the fact that Lee was the person who told us about this, but that isn't the main thing we pick up). Unlike local reductionism, the direct view has no problem at all with gaining knowledge from strangers: the classical Indian line is that knowledge can be gained directly not only from sages and 'noble compatriots' but also from 'barbarians', as long as they have knowledge and intend to share it.

Contemporary advocates of the direct view emphasize that trust in testimony plays a large role in the acquisition of language and in our everyday practices of communication. Where reductionists and Lockeans think it is right to maintain a neutral stance towards public testimony until we can verify it with our private resources (our own perceptions and inferences), advocates of the direct view suggest that we do not have sufficient private resources available to manage that kind of verification. Vital knowledge of what words mean, for example, is made possible only if we can gain knowledge directly by being told something by another. We wouldn't be able to understand each other in the first place if we didn't start by trusting others to tell the truth and accepting what they say at face value. On this view, we drink in what others say, in something like the way bees do.

Even for the maximally generous direct view of testimony, there are still certain conditions that must be met: in order to gain knowledge, what your informant says must actually be true, and (at least according to most non-reductionists) your informant must also know that it is true, as opposed to just managing a lucky guess. Plato gives an example to illustrate this last point: he describes a slick lawyer who has to defend a client against an assault charge. This client is in fact innocent, but has no witnesses to back him up. Although the lawyer himself has no idea whether what he is saying is true, he does a great job of telling the jury that his client is innocent, using his charisma to make the jury believe him. Plato then raises the question of whether the members of the jury actually *know* that the defendant is innocent, on the strength of what they have been told by the charming lawyer. The answer to this question seems to be 'no'. If you are going to gain knowledge of a fact from an informant, the informant had better know that fact himself.

One of the leading contemporary theorists of testimony, Jennifer Lackey, uses the image of a 'bucket brigade' to illustrate this 'take it from someone who knows' condition on testimonial knowledge: '[I]n order to give you a full bucket of water, I must have a full bucket of water to pass to you. Moreover, if I give you a full bucket of water, then—spills aside—the bucket of water you now possess as a result of our exchange will also be full.' The 'spills' here could include cases in which you don't quite hear what I say, or cases in which someone has maliciously told you that I am a pathological liar, and you doubt me for that reason. Both in the classical and in the contemporary versions of the direct view, gaining knowledge by testimony can be blocked if you have doubts about the truth of what is being said, even if the speaker does have knowledge, and even if your own doubts are unreasonable.

Lackey herself raises doubts about the idea that testimonial knowledge arises only when we take it from someone who knows. Imagine a schoolteacher who has personal doubts about a true scientific theory that she is required to teach to her class (say, a 'young earth' creationist who is teaching the theory of natural selection, according to which humans have an ancient evolutionary heritage, sharing common ancestors with other primates). This teacher doesn't know that the theory of natural selection is true—she doesn't even believe it—but nevertheless she diligently teaches it to her class because it is part of the state curriculum. If the teacher's trusting students come to believe the theory on the strength of their teacher's testimony, couldn't we say that these students now know the theory? If Lackey is right, this is a case in which someone with less than a full bucket manages to pass on more knowledge than she herself possesses.

There are other ways in which less-than-perfect sources might be able to pass on more than they individually know. One way is by working with others. Right now, I'm inclined to describe myself as knowing that the Willamette River flows northward between the Oregon Coast Range and the Cascade Range in the American northwest. I know this because I just looked it up in Wikipedia (go ahead, double-check). It's possible that someone who actually knows this fact was the one who first updated the page about this river to report it. Perhaps someone with a full bucket has poured their knowledge through the Internet to me. It's also possible that the person who first reported the watershed boundaries had something less than a full bucket, and was, say, slightly unsure of the name of the mountain range on one side. However, over time, the entry as a whole has been vetted by so many people that the line about those mountain ranges is by now well-secured by the whole community of editors. This group may have succeeded in filling the bucket together, jointly generating an entry that is now able to provide the reader with knowledge. If the reliability of an informant is what counts, groups working together under the right conditions can outperform single authors.

This example would be handled differently by advocates of the different theories we have covered so far. A Lockean would say that what we really take away from reading the Wikipedia entry on the Willamette River is just highly probable opinion, rather than knowledge, even if there are many, many informants who are telling us something entirely plausible. A reductionist would say that any knowledge we gain here is really inferential in character: on this view, the article could transmit knowledge to those who are already aware that Wikipedia entries are generally reliable. For example, the entry could succeed in transmitting knowledge about the river's course to those who have read the 2005 *Nature* article

7. The suffering of past generations

reporting that entries in Wikipedia are comparable in accuracy to entries in the *Encyclopaedia Britannica*, or to people who have significant personal experience of double-checking the accuracy of Wikipedia entries. Meanwhile, a direct theorist of testimony could say that Wikipedia provides knowledge of the facts it reports (when its internal systems of quality control are working well), even to those who are unaware of the reliability of those systems of control. On this view, even a naive 12-year-old preparing a school report could come to know the names of the ranges flanking the river, just by reading the Willamette River entry. Traditional versions of the direct theory would require whoever wrote or edited the key sentence in the entry to know the facts it reports; however, the direct theory could also be stretched or modified to allow cases in which we gain knowledge not from a single knower but from a largely anonymous community including individuals with partial confidence rather than full knowledge. As debate about the epistemic structure of testimony continues, new channels of information afford fresh opportunities for rival theories to offer competing explanations of the social transmission of knowledge.

Theories of knowledge generally turn to testimony only after they have examined perception and reason, but there are some philosophers who place it at the very heart of their approach to knowledge. Most notably, British philosopher Edward Craig argues that humanity came up with the concept of knowledge for the express purpose of managing the problem of testimony: we use this concept to mark people as good sources of information. Craig starts with the idea that all creatures struggling to survive need true beliefs about their environment. It helps us greatly if we are not restricted to what we have experienced personally but can also learn from others. It's imperative that we have a way of sorting out good informants, who can serve as our eyes and ears, from bad informants, who are likely to lead us astray. Good informants are identified as knowers.

Craig reverses the usual direction of explanation: most epistemologists think that being a knower is something that makes you (potentially) a good informant. Craig, however, considers the notion of 'good informant' to be more fundamental, and vital to explaining both the value and the evolutionary origin of the concept of knowledge. Critics of Craig emphasize that knowers can sometimes be bad informants—knowers can be secretive or deceptive. In addition, there are some intuitions about knowledge that Craig's theory has difficulty explaining. For example, it is hard for Craig to say why we see the victim of a Gettier case as failing to know. The victim of the Gettier case will in some sense be a good informant—as someone with a justified true belief, he is getting it right and in some sense thinking reliably. Meanwhile, the evolutionary dimension of Craig's theory is also open to question. Research on our closest animal relatives has shown that they can distinguish knowledge from ignorance, but not in the ways Craig's theory takes to be most basic. In experimental settings, chimpanzees can't distinguish between knowledgeable and ignorant informants who give them clues about where food is hidden. Chimpanzees can, however, keep track of who knows what when they are competing for resources: for example, subordinate chimpanzees are good at remembering whether a dominant animal knows where food is hidden. The connection between knowing and acting seems to be easier to spot than the connection between knowing and being a good informant. While testimony is an important topic in epistemology, it's doubtful that it will work as our starting point.

Chapter 7
Shifting standards?

Counting on context

Some words are slippery. Every night, the word 'tomorrow' slides forward to pick out a different day of the week. 'Here' designates a different place depending on where you are standing. 'I' stands for someone different depending on who is speaking; and 'this' could be anything at all. Words like 'big' and 'small' are also tricky: a morbidly obese mouse is in some sense big, but in another sense still small. What about the verb 'to know'? Is it possible that it also shifts around in some interesting way?

What the other words featured in the last paragraph have in common is context-sensitivity. The context in which these words are used plays a role in setting what they stand for. Some words (like 'I' and 'now') are sensitive to the speaker's identity and location in time and space. Others (like 'big' and 'tall') are sensitive to a comparison class: it takes much more height to be tall for a skyscraper than it does to be tall for a cereal box. To complicate matters, the same thing could be a member of two different classes (mouse, animal): a creature could be big for a mouse and still small for an animal, and which adjective we should use for it depends on which of these classes we have in mind.

It's tempting to say that context-sensitive words keep changing their meaning, but that's not exactly right. We don't have to buy a new dictionary every day to keep up on what the word 'tomorrow' means. There are some fixed rules: 'I' always picks out the person speaking and the relevant sense of 'tall' always means 'large for its kind in the vertical dimension'. Rather than changing their meanings, context-sensitive words work like recipes that take input from a conversational context to settle what they stand for. Once the context is established, it should be clear exactly what 'this' indicates, or which day of the week is picked out by 'yesterday'.

'Contextualism' is the standard name for the view that words like 'know' and 'realize' are context-sensitive. Some of the appeal of contextualism comes from its promise to reconcile the main points in Chapters 1 and 2 of this book. Chapter 1 observed that the verb 'to know' is one of our most common verbs, and is used as the default label for ordinary cases of seeing, hearing, or remembering that something is the case. (Of course, you know that you are reading a book right now.) Chapter 2 observed that it's easy to discover ourselves doubting that knowledge is ever humanly possible. (How could you ever really know that you are reading a book and not just dreaming that this is so?) According to contextualists, there's no real clash between the positive remarks in the first chapter and the negative remarks in the second. Because 'know' is context-sensitive, everyday claims about how much you know are fully compatible with sceptical claims about your knowing almost nothing. The everyday speaker and the sceptic are in different conversational contexts, and are therefore saying different things when they use the word 'know'. Just as the sentence 'Tomorrow is Friday' sometimes says something true, and sometimes says something false (depending on when you say it), knowledge-ascribing sentences like 'John Doe knows that he is reading a book' do a similar trick. When everyday low standards are in force, it's right to say 'John Doe knows'; when we are using the sceptic's high standards, the thing to say is 'John Doe does not know.'

The emergence of contextualism

Contextualism grew out of a theory of knowledge launched in the early 1970s, the 'Relevant Alternatives' theory of knowledge. Advocates of that theory say that knowing always involves grasping some kind of contrast. Here's an example: Jane Roe is at the zoo with her son, and sees a black-and-white striped animal in the enclosure ahead of her. 'Look, Billy!' she says. 'That's a zebra!' She's right; there's nothing wrong with Jane's eyesight or her ability to recognize ordinary zoo animals, and the black-and-white striped animal she is looking at is indeed a zebra. When asked whether Jane Roe knows that the animal she is looking at is a zebra, we find it easy to say 'yes'. But here's a trickier question: does she know that the animal she is looking at is not *a cleverly disguised donkey*? (Figure 8). It's possible to paint stripes on a donkey and trim its ears and tail, and from where Jane is standing, a cleverly disguised donkey would look exactly the same to her. Jane Roe's evidence, according to the Relevant Alternatives theory, is good enough to enable her to tell that the animal is a zebra (as opposed to a lion, antelope, or camel). Given the range of animals at an ordinary zoo, she has a pretty easy set of relevant alternatives to pick from. However, her evidence is not good enough to enable her to tell that the animal is not a cleverly disguised donkey: to make that harder judgement, she'd have to rule out the relevant alternative that it actually *is* a cleverly disguised donkey. It is not impossible for her to rule out that tricky alternative, but Jane would need to hop the fence and go closer, perhaps even close enough to dab some cleansing fluid on the animal's fur. So, from 20 paces away, she knows that the animal is a zebra, but she doesn't know that the animal is not a cleverly disguised donkey.

In this example, it's understood that being a cleverly disguised donkey is not a relevant alternative at an ordinary zoo like the one Jane Roe is visiting, but it could be a relevant alternative at an impoverished zoo, or at a zoo run by practical jokers. One of the challenges facing the Relevant Alternatives theory is coming up

8. A man stands with one of Marah Zoo's world famous painted donkeys, in Gaza City, Palestinian Territories, December 2009. Anastasia Taylor-Lind/VII

with the rules for setting what the relevant alternatives are for any given claim. The field of alternatives is partly set by what is actually possible for the speaker, but the further details of the theory have been hard to work out. Does it matter whether Jane Roe knows that the zoo is not run by jokers? It's unclear. But more importantly, there's something deeply *strange* about the Relevant Alternatives theory.

Let's assume that Jane knows some basic biology: suppose she knows, as most adults do, that zebras and donkeys are different species of animal, so that nothing can be both a zebra and a donkey

Box 6 A problem for the Relevant Alternatives theory of knowledge

What Jane knows from 20 paces away:

- That animal is a zebra.
- If that animal is a zebra, it is not a cleverly disguised donkey.

What she doesn't know:

- That animal is not a cleverly disguised donkey.

at the same time. If the animal is a zebra, then it follows from basic biology and logic that the animal is not a donkey. Logic also tells us that if something is not a donkey, it is not a cleverly disguised donkey. What the Relevant Alternatives theory says is that as Jane stands 20 paces away from the zebra, she knows that the animal she sees is a zebra, and also knows that if it is a zebra it is not a cleverly disguised donkey, but she doesn't know that the animal is not a cleverly disguised donkey. She can know the premises of the simple argument in Box 6 without being able to know the conclusion, despite being able to see for herself that it follows logically. The Relevant Alternatives theory therefore violates a principle called Closure, according to which you know anything that you succeed in deducing logically from your existing knowledge.

Violations of Closure seem weird: can't you trust logical deduction? Furthermore, as soon as we say that Jane doesn't know that the animal is not a cleverly disguised donkey, it sounds odd to insist she still knows it's a zebra. However, there is something compelling about the initial observations that motivated the Relevant Alternatives theory: it sounded right at first to describe Jane as knowing that the animal was a zebra, and it also sounded right to deny that she knew at a distance that the animal was not a donkey in disguise.

Contextualism emerged as a way of keeping the appealing part of the Relevant Alternatives Theory without accepting its strange denial of Closure. Published in 1976 by Gail Stine, the first clear formulation of contextualism was a two-part proposal. Stine's first claim was that we use higher or lower standards for knowledge in different settings: 'It is an essential characteristic of our concept of knowledge that tighter criteria are appropriate in different contexts. It is one thing in a street encounter, another in a classroom, another in a law court—and who is to say it cannot be another in a philosophical discussion?' Rather than having a field of alternatives fixed by what is possible for the person who is making the judgement, Stine proposed that various narrower or wider fields of alternatives are surveyed by those who are talking about whether a given person has knowledge. Stine's second point was to insist that within any given context, we appropriately stick to one set of standards. In a context in which we are worried about the possibility of doctored donkeys wearing make-up, it is wrong to say, 'Jane knows that the animal she is looking at is a zebra.' If she doesn't share our worries, Jane can truly say, 'I know that's a zebra,' but she can't then use pure logic and biology to deduce the exotic conclusion that it's not a cleverly disguised donkey. As soon as Jane starts thinking about cleverly disguised donkeys, her standards for knowledge will rise to take such exotic possibilities into account, and it will become wrong for her to say 'I know it's a zebra.' It's legitimate to talk about knowing with high standards or low standards; you just can't slide back and forth between the two in a single context without marking the shift.

In Stine's formulation of contextualism, knowing requires discrimination from a larger or smaller field of alternatives. The everyday speaker wonders whether the sandwich is chicken or tuna, and counts a person as knowing if he can rule one of those options out in favour of the other; the sceptic wonders about many more exotic alternatives as well (hypothetical new kinds of tuna that look like chicken, mere holograms of sandwiches, projections sent by an evil demon, and so forth). Once we have this wider array of alternatives to contend with, it's harder to credit anyone with knowledge. The widening and narrowing of a field of relevant alternatives is one vivid way in which contextualism can be expressed. But the basic idea behind contextualism is something more general, and not necessarily married to the relevant alternatives idea: the idea is that 'know' expresses something different as situations change, and different contextualist theories have different lines about how this works. There are internalist formulations of contextualism in which different contexts call for more or less evidence, and externalist formulations of contextualism in which belief-forming processes must track the truth across a narrower or broader array of circumstances as standards rise or fall. Contextualism on its own is not a theory of knowledge: it's a theory about knowledge-attributing language, a semantic topping that can be spread onto various different underlying theories of knowledge.

Is it an appetizing topping? Contextualism does promise a neat solution to the problem of scepticism: both sides are right, in a sense. The man in the street who says, 'I know that I am reading a book' is saying something true, but so is the sceptic who says, 'The man in the street does not know that he is holding a book.' The trick is that the word 'know' picks out a different relation for each of these speakers. Talking about scepticism raises the standards for knowledge, just as talking about basketball players raises the standards for 'tall', to use a favourite contextualist analogy. A man who measures six feet (183 cm) ordinarily counts as tall in the United States of America, where the average male height is about 5'9" (175 cm), even if six feet doesn't count as tall in the National Basketball Association, where the average height is about 6'7" (201 cm). Consider Chris Paul, a six-foot player for the LA Clippers. Sports fans discussing the strengths and weaknesses of the team can honestly say 'Chris Paul isn't tall' at the very moment that Chris Paul is honestly describing himself as tall on his dating website profile. There is no real conflict here, thanks to the difference in standards.

We have to be careful about speakers who are talking about what other speakers say: it would be a mistake for Chris to say 'What those sports fans said about me was false' simply on the basis of his being above average in height for an American man; it would also be a mistake for the fans to accuse Chris of saying something false on his dating profile simply on the basis of his being short for a professional basketball player. When you judge the truth or falsity of other speakers' context-sensitive remarks, you need to respect their context (and they need to do the same for you). The most straightforward thing to do would be to make the comparison class explicit. The sentence 'Chris Paul is tall' leaves a blank (compared to what?) that has to be filled in by context; both of the sentences 'Chris Paul is tall for an American man' and 'Chris Paul is not tall for a professional basketball player' have that blank filled in, and so express truths in a stable fashion across both contexts.

Going back to the case of knowledge, what's the clearest thing to say? Can we make our standards explicit? Is 'You know that you are reading a book' just like 'Chris Paul is tall'? If so, we should just be able to solve the problem of scepticism by saying something like 'You know, by low standards, that you are reading a book, but you don't know, by high standards, that you are reading a book.' Well, that was easy. But was it satisfying?

It's hard to say. Defending the context-sensitivity of 'know' isn't quite as smooth as defending the context-sensitivity of 'tall'. One objection that is often raised against contextualism is that it is really just scepticism in disguise. A sceptic doesn't have to fight the low-standards notion of knowledge in play in ordinary life; sceptics themselves doubtless use the verb 'know' dozens of times every day when they are talking to each other about ordinary things. What the sceptic really wants to say is that when we look closely and carefully at knowledge, we'll see that our everyday claims to have it are actually false. This doesn't amount to a call for stamping out all talk of knowledge, because we may succeed in communicating something useful in our casual talk about knowing. It could be interesting to hear a co-worker say, 'Lee knows who got the job,' for example. Thinking about the strict meaning of 'know', we may decide that Lee doesn't really know who got the job (could Lee really guarantee that he is right, and that the apparently successful candidate hasn't just been struck by a meteorite?). Maybe by saying 'Lee knows,' what's really communicated is that it's a good bet Lee could tell you. It's not crazy to think that you can convey something useful by saying something literally false: you use many words in this loose manner, like saying 'I'm starving,' when you are really just hungry. Even if that's literally false, it can help to get your host to offer you a snack. In the sceptic's high-standards context, 'Lee knows who got the job' expresses something false, even if we do something useful by saying that sentence in everyday contexts. But if contextualism says that sceptics speak truly when they deny that ordinary people know simple facts (like the fact that you are reading this book), then contextualism seems to be letting scepticism win, at the expense of common sense.

Contextualism is actually more subtle than the rough 'loose use' theory sketched above. In particular, contextualists are careful about respecting the context of other speakers. According to contextualism, sceptics get to say, truly, 'Lee doesn't know who got the job.' But they *don't* get to say that everyday speakers are saying something false when those everyday speakers say, 'Lee knows who got the job' in an everyday context. The same reasoning applies on the other side: ordinary people can express something true by saying, 'I know that I am reading a book,' in an ordinary context, but they cannot say that the sceptics are speaking falsely in saying, 'You don't know that you are reading a book' in the context of a high-standards philosophical discussion. Contextualists themselves don't have to say that the sceptical philosopher's context is better (or worse) than the common person's: high standards are not necessarily better than low standards (in fact, high standards can be really annoying and pedantic when you just want to send a card congratulating the person who got the job). Both sides win, as long as they play nice and refrain from putting down what the other side is saying.

Contextualism's tender tolerance for other points of view does not appeal to everyone. Critics of contextualism continue to resist the idea that the sceptic and the common man are both saying something true, and continue to wonder which way of talking *really* manages to get it right about knowledge itself, once and for all. From a contextualist perspective, asking which way of talking ultimately captures the nature of knowledge is like asking which weekday is ultimately 'tomorrow'. It's not a good question.

If contextualism aims to be somewhat friendly to both the sceptic and the common person, it has to be less friendly to the philosophical adversary who thinks there is a single true answer to the question 'Does Lee know who will get the job?' no matter which context we ask it in. That person really is making a mistake, contextualists will say, and they will say the same about anyone who thinks that the common person and the sceptic can't both speak truly. Contextualists do recognize that it's very common to think that one is forced to take sides: somehow the way in which 'know' is supposed to shift is more hidden from us than the way in which 'tall' or 'here' is. We don't have a big philosophical tradition of debates about which place is *really* 'here', in the way that we have debates about which way of talking is really right about knowledge. But why exactly would the context-sensitive workings of our language be obscured to us when we are talking about knowledge, if they are so transparent when we are talking about times, places, and qualities like 'tall'? One of the most active current research questions for contextualists concerns just this question, and various proposals have been advanced. Perhaps something about our use of 'know', like its role in closing off further enquiry, hampers us from tracking context shifts as well as we should, and gives us an illusion that knowledge is absolute. Or perhaps contextualism is wrong, and knowledge itself really is absolute. The view that knowledge is absolute, in the sense that the words we use for it are not context-sensitive, is known as 'invariantism'. Invariantism faces a challenge in explaining the shifting intuitions that make knowledge sometimes seem easy and sometimes seem hard.

Interest-relative invariantism

Lee is walking towards the bus stop after work when he bumps into his co-worker Smith, who is headed back into the building.

> —'Lee, do you know if the ground-floor supply room door is locked? I just realized I left my jacket in there, and I don't have a key.'
>
> —'Yes, it is—I know because I locked it myself half an hour ago, and I didn't see anyone else in that hallway after. Sorry!'

Lee doesn't seem to be saying anything controversial in claiming knowledge here, and we'll assume as we go forward that in fact the door really is locked, and that Lee's key, his eyesight, and his memory are all fine. But now we'll imagine a different version of the story, one that takes a different turn after Lee leaves the building. As Lee walks towards the bus stop, he is approached not by Smith, but by four police officers.

> —'Excuse me, sir, but we have an emergency situation in the building you just left. Apparently there was a shooting on the second floor, and the gunman is still in the building. Are there any ways out of the building other than the front door here?'
>
> —'There's a back door leading off the supply room, but I locked the door to that room half an hour ago.'
>
> —'Do you know if it's still locked, or if anyone else might have opened it?'
>
> —'I don't know—I didn't notice anyone go by, but I wasn't watching the door the whole time.'

In the forgotten-jacket version of the story, Lee claimed to know that the door was locked; in the gunman version, he claimed not to know. Both times he was saying something that sounded true. The curious thing is that both stories ran parallel up to the moment Lee left the building: in both stories he is trusting his memory of the last half hour as he answers the question about whether he knows. In traditional epistemology, whether or not you know depends on traditional factors such as whether your belief is true and how good your evidence is: interestingly enough, all these factors seem to be the same in both stories. So how is it possible that Lee knows the door is locked in the first, but doesn't know in the second?

It's clear what contextualists would say: in the casual bus-stop conversation, there are low standards in play, and when the police get involved, the standards rise. Lee is saying something true when he says 'I know' in the first story, but also when says 'I don't know' in the second. But this is not actually a story about how Lee knows in the first story and not in the second: it's a story about what a person can say truly in the two contexts. Viewed from other ('higher') perspectives, it would be *false* to say that Lee knew the door was locked in the first story, or so contextualism maintains. There is no simple, context-independent answer to the question 'Did Lee know or didn't he?'

If we were pushed towards contextualism by the feeling that it seems like Lee really does know in the first story, and doesn't in the second, then we might try to find a theory of knowledge which would fit those feelings more directly. If all the traditional factors that matter to epistemology (truth, evidence, reliability, and so forth) are the same in the two stories, one possibility is to allow some non-traditional factors to make a difference. What other factors are different in the two stories? Advocates of a position now known as interest-relative invariantism (IRI) have noticed that there are *practical* differences between the two stories. In the first, not much is at stake for Lee; as a friendly colleague he'll go back and unlock the door so Smith can get his jacket, but if he's wrong about the door still being locked, it's not a big problem. Finding it open would mean only that he'd wasted a minute on a short walk. In the second story, there could be serious practical consequences if Lee is wrong about the door being locked: the gunman could escape out the back exit.

Interest-relative invariantists have found that practical interests seem to have an impact in many cases. Do you know whether the sandwich is chicken or tuna? If not much is at stake (you like chicken just a little bit more, and you would slightly prefer to get soup instead if it's tuna), then a fairly casual inspection (looks like chicken) would be enough to count as knowing. If your life is at stake (you have a very serious fish allergy), then you will not know on the basis of that causal inspection, according to IRI. The more you have at stake, the more evidence you need in order to count as knowing. This is how the two versions of the Lee story can deliver different verdicts about whether Lee knows.

What makes IRI different from contextualism is that IRI is a theory about how knowledge itself works, not just a theory about the semantics of knowledge-ascribing vocabulary. The verdicts that it delivers are not just true-when-expressed-in-certain-contexts, but true full stop. Lee knows in the first story; he doesn't know in the second. The context that matters in setting the standards for how much evidence Lee needs for knowledge is Lee's own context, not the contexts of other people who might be talking about him from different perspectives. According to IRI, the sceptic is just wrong to say that Lee lacks knowledge in the first story. Different advocates of IRI have different accounts of the nature of knowledge; what they all have in common is that practical interests, which are not a factor for traditional epistemologists, are a factor that helps determine whether or not a person knows: as stakes rise, more evidence is needed for knowledge. Advocates of the theory contend that IRI is the best way of making sense of the relationship between knowledge and action.

Old-fashioned invariantism, again?

Contextualists were quick to criticize the IRI approach, noticing, for example, that it also has trouble capturing some patterns of shifting intuitions. If it's a plain context-independent fact that Lee knows that the door is locked in the first (forgotten-jacket) version of the story, how is it that the sceptic is able to make us start doubting that fact so easily? The IRI approach also gets awkward when we talk about counterfactual possibilities. 'The waiter doesn't know whether the sandwiches are tuna or chicken, but he would have known if it weren't for the fact that one of his customers has an allergy.' That sounds odd, but IRI predicts it shouldn't. Meanwhile, advocates of IRI have fired back at the contextualists, often by pointing out that 'know' doesn't really work like other context-sensitive vocabulary: unlike 'tall', it doesn't easily fit on a sliding scale, and unlike 'today', there are no simple rules for explaining how context fills it in. Contextualists have suggested that 'know' might have a special kind of context sensitivity all its own, but it's still an open question exactly how it works.

As the main shifting-standard views take shots at each other, advocates of more rigid standards have wondered whether their old-fashioned view might still win the day. Strict invariantists maintain that knowledge is determined strictly by traditional factors (truth, evidence, and the like) and that knowledge-attributing vocabulary is not sensitive to context. We've already met one of the most straightforward forms of strict invariantism: scepticism. According to Academic Scepticism, for example, there is a single fixed standard that must be met in order to have knowledge (we must have an infallibly correct impression of the thing judged). Sadly, we never meet this standard in daily life (or perhaps we meet it only for one or two special claims, like 'I exist'). If you take yourself to know you are reading a book, you are just wrong. Sceptics seem to owe us a story about why we speak so much of knowledge if it is forever out of reach, and we may or may not be satisfied with what they have to say. If we are unsatisfied, we may want to turn in the direction of moderate strict invariantism, which holds that there is a single fixed standard that must be met in order to have knowledge, but also that it is a standard that humans often meet. You do know that you are reading a book, and Lee really did know that the door was locked. Strict moderate invariantists also owe us something: they need to explain why exactly the sceptic can so easily lead us to doubt our everyday judgements, and why Lee's knowledge seemed to melt away as he was questioned by the police, even though the strictly traditional factors he was relying on remained the same. Strict moderate invariantists have struggled to answer these questions. One avenue they have tried is to argue that there is something wrong with our shifting intuitions, or with the cases that produce them. Perhaps the differences between the two versions of the locked-door story are larger than they seem: we assumed that the traditional factors that matter to knowledge were identical across those cases, but it's possible that the high stakes in one case will naturally trigger lower confidence or a different way of thinking about one's evidence. Or perhaps something about the high-stakes situation makes us confused about the difference between knowing and knowing that we know, or about the difference between what we are literally saying and what we are trying to convey. Perhaps there is something wrong with our instincts about these cases; perhaps some natural distortion is introduced when we talk to the sceptic or weigh life-and-death issues. Given how hard it is to develop a smooth story about our patterns of intuition about knowing, it makes sense to take a deeper look at how those intuitions are produced.

Chapter 8
Knowing about knowing

Epistemology's raw materials

How much do we know about knowledge before we start to study it systematically? We don't seem to start completely empty-handed. Philosophers have a special reason to hope that we have something at the outset, some talent for spotting genuine cases of knowledge: instincts or intuitions about particular cases are supposed to support some philosophical theories over others. If you felt that the person looking at the broken clock didn't know the time, that feeling was taken to work as a reason for you to reject the classical analysis of knowledge. But what enabled you to judge that case the way you did? And can you tell whether your way of judging—whether or not others share it—is really getting it right? These questions have recently sparked fresh empirical and philosophical work on our intuitions about knowledge.

Reading minds

The word 'intuition' may suggest some mystical power of insight, but intuitions about knowledge are a feature of everyday life. Focus for a moment on the difference between these two claims: (1) 'Lee thinks that he is being followed.' (2) 'Lee knows that he is being followed.' There's a significant difference here, but the choice to say one thing rather than the other doesn't usually involve deliberate calculation. You can feel that someone knows something (or doesn't know it) without calling to mind any explicit theory of knowledge: this kind of feeling is an intuition. There's a name for the natural capacity that generates instinctive feelings about knowledge and other mental states: mindreading. The word is popularly used for stage magic acts in which a magician performs an impossible-seeming trick of reading another person's thoughts. As psychologists use the term, it applies to something that we do a hundred times a day with no special fanfare. Mindreading is the attribution of 'hidden' or underlying mental states—wanting, fearing, thinking, knowing, hoping, and the rest—to another person. When you see someone reach towards something, you do not simply have the impression that an arm is extending in space; you see the person as reaching for the salt, wanting something, and aiming to get it. As we register, perhaps only subconsciously, which way people are reaching and glancing, and as we detect tiny shifts in their facial expressions, we gain a sense of what is going on inside them and how well they grasp their environment, and we become better able to predict how they will act and interact with us. Without a capacity for mindreading, we'd be stuck looking at surface patterns of moving limbs and facial features; mindreading gives us access to deeper states within a person. Whether we are trying to coordinate or compete with others, it helps enormously to know what they want and know, and whether they are friendly, angry, or impatient. We don't always get it right: it's possible to mistake what someone knows or wants, or to be misled by a skilled deceiver, but our daily social navigation is so effective that it comes as a surprise when we occasionally misread a situation.

The mindreading abilities of human beings are better than those of any other species on earth. Chimpanzees monitor whether their rivals do or do not know about hidden food: they grasp the simple contrast between having and lacking knowledge. But human beings can also keep track of the ways in which others are mistaken, and this is something that no other animal does (as far as we can tell). You can see another person as having a false belief: playing a practical joke on a friend, you empty his cereal box and fill it with rubber insects. You know, as you watch him sitting down at the breakfast table, that he is expecting cereal to pour from the box. You know that he has an inner representation that doesn't match the actual outer reality. No other animal seems to be capable of representing this kind of state, even in situations in which it would be highly advantageous. In experimental tests of whether an animal can keep track of another's false belief, all non-human animals fail.

These tests are hard enough that human beings fail too, at least when they are toddlers. Here's one of the classic demonstrations of this failure: a young child is given a familiar container, such as a candy box with a picture of candy on it, and asked what is inside. ('Candy!') Then there's a surprise: the box is opened to reveal that it contains not candy, but crayons. The box is closed again, and the child is then asked what the next child, waiting outside, will think when she comes into the room. Will that child know what is in the closed box? What will she think is in the box? Almost all five-year-olds recognize that the next child will have a false belief about the contents of the box, but only a small minority of three-year-olds can get this right. Strangely enough, most three-year-olds say that the next child will already know that there are crayons in the box, and even more strangely, many three-year-olds give the wrong answer when they are asked what they themselves thought at first on seeing the box. Calculating another person's false belief is a demanding task, as is remembering your own past mistake: you have to suppress your current picture of how the world is in order to depict things from the mistaken person's point of view. Belief or opinion, a state which may or may not match the outer world, is a relatively tricky state to represent. If knowledge is a state that must reflect how things are, then the attribution of knowledge is simpler.

Advocates of the 'knowledge first' programme in epistemology have claimed support for their position from these findings and from various other lines of research supporting the notion that states of knowledge are easier to represent than states of mere belief. For example, across cultures, the verb 'know' is acquired earlier and used more heavily than the verb 'think'. Critics have argued that the order in which we learn these concepts may not reveal which of them is really basic: after all, we learn about table salt well before we learn about the elements that make it up, and we also talk about the compound more often. Critics have also raised questions about whether children too young to grasp the concept of false belief can already be credited with the concept of knowledge. To complicate matters further, recent research suggests that young children and even infants do have some partial and implicit capacity to recognize false beliefs, although the scope and significance of this capacity are still unclear. We are still struggling to figure out the true developmental story about the human ability to spot knowledge, but continuing research into this difficult territory can still give us a clearer picture of the natural structure of our everyday concepts of knowledge and belief.

More broadly, empirical work can also give us a better sense of the natural limits of our powers of mindreading. These powers involve some very specialized equipment. You can flex this equipment by reading the three very short stories in Box 7, used in research led by MIT neuroscientist Rebecca Saxe.

As you were reading these stories, different regions of your brain were selectively activated. Young children's brains don't respond very differently to the second and third stories, but for older children and adults, the third story is quite distinctive. To understand it, you need to represent the characters' states of mind. Reading stories like this selectively activates a brain region near your right ear: the right temporo-parietal junction (RTPJ). Other regions are involved as well, including the medial prefrontal cortex (MPFC), which is activated in adults and children by all kinds of stories about people, including the second story above. The RTPJ is also activated in young children for all stories with social content,

Box 7 Stories that activate different brain regions

1. Physical

Out behind the big red barn at the edge of the walnut grove is the most magnificent pond in the neighbourhood. It is wide and deep, and shaded by an old oak tree. There are all sorts of things in that pond: fish and old shoes and lost toys and tricycles, and many other surprises.

2. People

Old Mr McFeeglebee is a grey wrinkled old farmer, who wears grey wrinkled old overalls, and grey wrinkled old boots. He has lived on this land his whole life, longer even than most of the trees. Little Georgie is Mr McFeeglebee's nephew from town.

3. Mental

Mr McFeeglebee doesn't want any little boys to fish in the ponds. But little Georgie pretends not to notice. He likes fishing so much, and besides, he knows he can run faster than anybody in town. Georgie decides to run away really fast if Mr McFeeglebee sees him fishing.

but the mature RTPJ responds in particular to stories about what people know, want, decide, pretend, believe, and so forth. When this section of the brain is damaged by a stroke or lesion (or temporarily disabled in an experiment, through transcranial magnetic stimulation), we become deeply impaired in our ability to evaluate and predict what others know and what they will do.

It's not surprising that mindreading comes to have its own specialized area in the adult brain: although we can do it effortlessly, mindreading involves some complex calculations. In this respect it has something in common with face recognition, which also involves very rapid and effortless calculations, and is also highly specialized within one specific area of the adult brain. What is the relationship between what someone wants, notices, pretends, and plans to do when he is seen? Navigating around these patterns is a non-trivial task. If we are able to decide, in the course of a casual conversation, whether to describe someone as just thinking or really knowing something, we can do this without conscious calculation in part because we have specialized brain resources devoted to the task of tracking mental states.

There are natural limitations to our mindreading equipment. One limit is a simple capacity limit on how many nested levels of mental state we can represent. Here's a sentence which uses four levels: 'Davis thinks that Lee knows that Smith doesn't want Jones to find out about the job.' Did you follow that? Researchers have tested people on their capacity to track up to nine levels, but most adults can manage only five levels before they break down and start to answer comprehension questions randomly (people with larger social circles do a bit better). This limitation is like our limitation in tracking moving objects on a screen (most people can manage five, heavy video gamers can go a bit higher): we have only so much attention to go around.

Here's a deeper limitation that is specific to mindreading: we have a tendency to be self-centred. More precisely, we suffer from a bias called 'egocentrism', which makes it difficult for us to override our own perspective when we are evaluating others who know less about their situation than we do. The bias is very pronounced in young children: the candy-box task described above is one example. A young child who knows that there are crayons in the box has trouble imagining that others could fail to know this. Children also have trouble recognizing the limits of others' informational access to the world. For example, having lifted a pair of visually identical piggy banks, one light and one heavy, even older children able to pass the candy-box task will wrongly predict that someone who just sees the two piggy banks from a distance will be able to tell which is the heavy one.

Even adults still make errors along the same lines. It remains difficult to bracket our own private knowledge of a situation when calculating the perspective of someone more naive. For example, in stock-trading games where one side is given special 'insider' information that they know the other side lacks, the better-informed side has trouble setting aside what they know in calculating how the other side will respond to offers, even when it would be really advantageous to calculate the naive view accurately. It's also tough to suppress our own private knowledge of a situation when we are evaluating how well or poorly someone has been thinking in making a decision: if you are told that a long-odds gamble ended up working out favourably, that information will colour your assessment of the wisdom of the gambler to a much larger extent than you might expect, even if you recognize that it is unreasonable on your part to take that outcome information into account.

It's not entirely clear why we have so much trouble subtracting from our own special knowledge when trying to represent or evaluate other perspectives, but we do know that the egocentric bias is very robust: some biases can be suppressed if you forewarn people about them, or if you give cash incentives for better performance, but egocentrism sticks with us even under those conditions. Epistemologists have wondered whether this limitation on our natural capacity to see other perspectives could be playing some role in the pattern of intuitions motivating contextualism. Once I am thinking about tricky alternatives like disguised donkeys, I will have trouble evaluating the perspective of a naive zoo visitor: even if I'm explicitly aware that she isn't thinking about those strange possibilities, egocentrism can drive me to evaluate her as though she is. Whether or not this works as a strategy for explaining the intuitions behind contextualism, epistemologists can profit from a better understanding of the natural mechanisms behind our intuitions about the presence and absence of knowledge. If some intuitions can be shown to arise from natural limitations or biases in our mindreading capacity, we can handle them with special caution as we construct our theories of knowledge.

Challenges to the case method

Philosophers interested in the nature of our intuitions can look at empirical work on mindreading, but they can also try something much more direct: polling ordinary people about their intuitions on the cases that matter to epistemology. Known as experimental philosophy, this method turned up some surprising results, especially in its early days. One of the first published papers in experimental philosophy, a 2001 paper by Jonathan Weinberg and colleagues, investigated responses to a series of vignettes about people making judgements. Some of the cases were easy: does a person who has a 'special feeling' that a coin will land heads actually know that it will? (over 90 per cent said 'no'). Some of the

Box 8 Weinberg and Colleagues' American Car case (2001)

Bob has a friend, Jill, who has driven a Buick for many years. Bob therefore thinks that Jill drives an American car. He is not aware, however, that her Buick has recently been stolen, and he is also not aware that Jill has replaced it with a Pontiac, which is a different kind of American car. Does Bob really know that Jill drives an American car, or does he only believe it?

Circle one:

REALLY KNOWS   ONLY BELIEVES

cases were hard, involving complicated hypothetical conspiracies. One of the harder cases (featured in Box 8) was a Gettier-type story. An overall majority of the experimental participants reported that the central character in the case lacked knowledge, judging the case in the same way that mainstream philosophers would. But not every subgroup performed the same way: while 74 per cent of the 66 participants who self-identified as 'Western' gave the standard answer, only 43 per cent of the 23 self-identified 'East Asian' participants and 39 per cent of the 23 'Indian Subcontinental' participants did so. Weinberg and colleagues concluded that philosophers who trust their intuitions in epistemology are not discovering the objective nature of knowledge but rather are engaged in the quite different activity of revealing their own local cultural attitudes, attitudes which might not be shared by other groups, and need not show us anything about the nature of knowledge itself. The experimental philosophy movement advocated abandoning the use of intuitions in philosophy.

Traditional philosophers were annoyed. Some of them objected to details of the case: its subject matter—American car brands—could interest some groups more than others, and the story is more readable to those who recognize at once that a Buick is an American car. Furthermore, the story is open to being fleshed out in different ways: are we supposed to think of Jill as the kind of person who just always drives an American car, and could her friend Bob know this fact about her? If we think of the story that way, it's not even a Gettier case: the theft of the Buick doesn't destroy the long-standing tendency of Jill's that is known to her friend. Other philosophers argued that we can't place too much stock in the quick responses of amateurs, which are doubtless influenced by all kinds of random factors: intuitions about philosophical scenarios only count if they are the products of careful reflection. Caution is needed in pushing that last line, however, at least among philosophers who claim that they are developing theories to make sense of pre-theoretical intuitions. If you have to think really hard about what your intuition is on a case, critics charged, your theory might be contaminating your judgement.

Later research challenged Weinberg's original findings while leaving some of experimental philosophy's larger questions in place. More systematic investigations of a wider variety of cases have failed to show that there is anything peculiarly Western about the intuitions traditionally reported by philosophers. Larger studies involving many participants from different cultures have in fact shown quite consistent patterns of knowledge attribution, across men and women of different cultures, for various Gettier cases, including the original 'American Car' story. Ordinary people do tend to agree with professional philosophers in judging that the agents in Gettier-type cases lack knowledge. Where Weinberg and colleagues were concerned that Gettier responses arose only from a very particular cultural group, and couldn't therefore be taken as evidence about knowledge itself, subsequent work favours the notion that these responses are universal. These more recent findings fit with cross-cultural work on the development of the concept of knowledge. Children in large urban pre-schools in London and Toyko go through the same broad sequence of stages in separating knowledge from ignorance and from false belief as children in rural hunter-gatherer societies in Cameroon. To be sure, adult performance is sometimes different across cultures—for example, Chinese adults seem to outperform American adults in the speed and accuracy of certain computations of other points of view—but the underlying competence to spot knowledge seems to be the same across cultures. If early work in experimental philosophy raised the worry that philosophers had peculiar intuitions about knowledge not shared by others, more recent work seems to be putting those worries to rest. Deeper worries remain, however. The fact that an impression is broadly shared is not always a sign that this impression is right.

Just as ordinary people share professional philosophers' intuitions about Gettier cases, they share their intuitions about some more problematic cases. Imagine a man called Albert in an ordinary furniture store. Albert sees a bright red table that he likes, checks the price, and asks his wife, 'Do you like this red table?' Thinking about the case, would you say that Albert knows that the table is red? (Of students polled in a recent study, 92 per cent said 'yes'.) Now consider a version of the story that just adds a few more details. The table is indeed red, and the lighting in the store is normal. But we will draw your attention to a further fact: if it were a white table under a bright red spotlight, it would look *exactly the same* to Albert, and Albert *has not checked whether the lighting is in fact normal.* Does Albert know that the table is red? (Suddenly less than half of the survey participants say 'yes'.) For some reason, thinking about a possibility of error, even when it's clear that Albert is not at risk of making that error, greatly reduces people's inclination to see Albert as knowing. Notice that the two stories weren't about two characters in objectively different settings; the second story just added some details that could also have been stated in the first version as well. It's as if we shifted from being generous about knowledge to being stingy when we thought about the case harder. This is the pattern of intuitions that has driven some philosophers towards scepticism and pushed others towards contextualism.

On the surface, there is an inconsistency between the two judgements. Sceptics think that only our second intuition (that Albert lacks knowledge) can be trusted; after all, that's the intuition we had when we thought harder about the situation. Contextualists think that the apparent inconsistency isn't a real inconsistency: both intuitions are fine, and they don't even really clash, because we meant something tougher by 'knows' in the second case, when we were thinking about tricky possibilities. Albert met the low standard but not the high one, just as a man could be tall for an American but not tall for a professional basketball player. Meanwhile, moderate invariantists think that

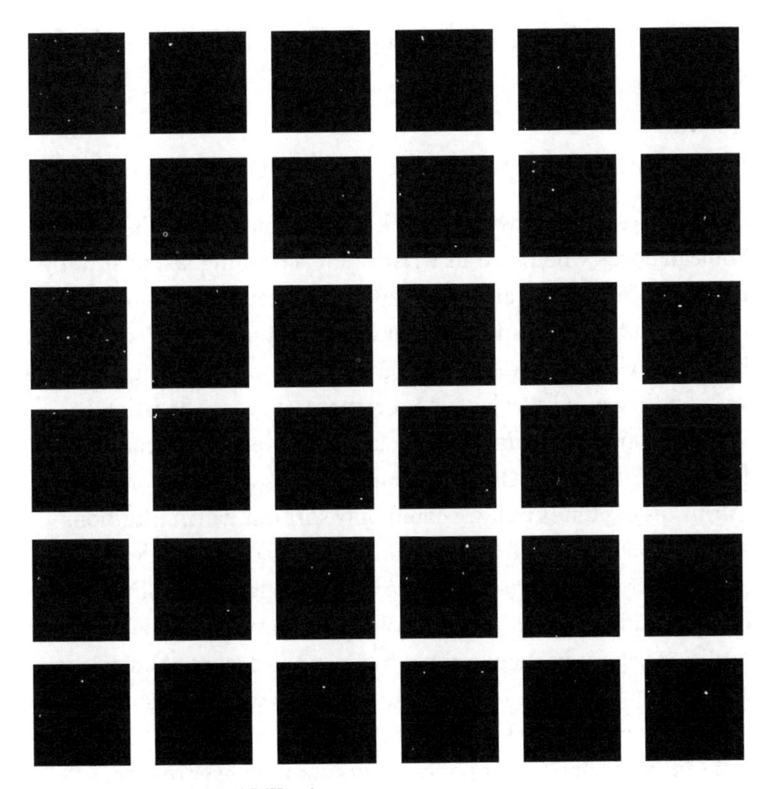

9. The Hermann Grid illusion

the second, more sceptical, intuition is the one that's mistaken, perhaps arising from some psychological bias. For example, we could be evaluating Albert as if he himself is thinking about the possibility of tricky lighting but perversely refusing to check whether the lighting is normal before he makes up his mind that the table is red. According to the moderate invariantist view, our response to this case is a cognitive illusion: something in the brain naturally miscalculates here, the same way the eye naturally miscalculates the shading of the spots at the corners of the boxes in the Hermann Grid illusion (Figure 9). Work in experimental philosophy now shows that this pattern of intuitions is widely shared, but it still doesn't tell us which of the intuitions in the pattern are really capturing the nature of knowledge, and which intuitions, if any, are illusory. How can we resolve this problem?

Impressions of knowledge, and knowledge itself

Some philosophers have argued that our situation here is hopeless. You can check whether your wristwatch is running fast or slow by comparing it to the National Research Council atomic clock, but there's no obvious counterpart for checking the accuracy of your intuitions about knowledge. If you've had trouble coming up with a smooth theory of knowledge that explains all of your instinctive feelings about particular cases, you may suspect that some of these feelings are illusions. But which ones? Philosopher Robert Cummins contends that we could only sort out which intuitions about knowledge were the right ones if we had some independent intuition-free access to the nature of knowledge itself. But if we had that kind of direct access we wouldn't have to fumble with intuitions about particular cases to grasp the nature of knowledge: 'If you know enough to start fixing problems with philosophical intuition, you already know enough to get along without it.' Cummins concludes that philosophers should never rely on intuitions about cases in formulating theories of knowledge.

That's a very pessimistic response. The parallel move in the perceptual domain would be to say that we can't start fixing problems with our visual impressions of colour, and figuring out which ones are illusory, until we have an independent, vision-free way of accessing colour. Although we are starting to develop technologies that can sort out colour signals without reliance on the human eye—photometers can measure the colours on that diagram of the Hermann Grid and won't get confused about the corner spots—we didn't actually wait until we had that equipment to start sorting out the illusions from the accurate impressions. We have used a variety of techniques over time to figure out which impressions are right, including double-checking impressions in different contexts or from different angles. Our understanding of

10. Can we make progress?

vision has evolved alongside our understanding of the nature of light and colour.

Meanwhile, even without the epistemological equivalent of a photometer to detect states of knowledge, our investigation of intuitions about knowledge can also evolve alongside our investigation of knowledge itself. As we move towards a better understanding of our knowledge-spotting instincts, fitting them into a broader picture of our psychology, we become better able to sort out which intuitions should count. Meanwhile, a sharper philosophical picture of knowledge itself will help advance our understanding of the nature of those instincts. Tools other than intuitions can also be used to tackle the problem of knowledge. We can work to develop internally consistent theories of knowledge that fit with our broader theories of human language, logic, science, and learning. We can refine our rough intuitive sense of the conditions that make knowledge possible by building mathematical models of individual and group knowledge. We can compare the strengths of existing philosophical theories generated in a range of historical periods and across different cultures. Some philosophical claims about knowledge have turned out to be confused or self-undermining, but other findings about knowledge, like its special connection with truth, have stood the test of time. If we do not know in advance which of our methods will be best suited to deliver further insight into the nature of knowledge, this is in part because we still do not fully understand what knowledge is. But even if we have yet to secure full knowledge of the nature of knowledge, we now find ourselves in a better position to make progress on this ancient question.

References

Chapter 1: Introduction

Data on the frequency of 'know' and 'think' in English are from *Word Frequencies in Written and Spoken English*, by Geoffrey Leech, Paul Rayson, and Andrew Wilson (New York: Routledge, 2001). For other languages, see the Linguistic Data Consortium website (<http://www.ldc.upenn.edu/>).

The claim that 'know' is a linguistic universal is defended by Anna Wierzbicka in her *Semantics: Primes and Universals* (New York: Oxford University Press, 1996).

The observations about verbs of ingestion and motion, and the examples in Box 1, are from Cliff Goddard's 'Universals and Variation in the Lexicon of Mental State Concepts', in *Words and the Mind: How Words Capture Human Experience* (New York: Oxford University Press, 2010).

The classic treatment of the factivity of 'knows' is Paul Kiparsky and Carol Kiparsky's paper 'Fact', in M. Bierwisch and K. Heidolph (eds), *Progress in Linguistics* (The Hague: Mouton, 1970). There is also a helpful discussion of factivity in the first chapter of Timothy Williamson's *Knowledge and its Limits* (Oxford: Oxford University Press, 2000).

The explanation of the projected use of 'knows' is drawn from Richard Holton's paper 'Some Telling Examples', *Journal of Pragmatics*, 28 (1997): 625–8.

Protagoras's relativism is discussed in Plato's dialogue *Theaetetus*. *The Theaetetus of Plato*. Trans. M. J. Levett, ed. Myles Burnyeat (Indianapolis: Hackett Publishing Company, 1990).

Chapter 2: Scepticism

The quotations from Sextus are from pp. 48–9 of Julia Annas and Jonathan Barnes's edition of *Sextus Empiricus: The Outlines of Scepticism* (Cambridge: Cambridge University Press, 2000).

The quotation from Śrīharśa's *The Sweets of Refutation* is from Ganganatha Jha's translation, 2nd edition (Delhi: Sri Satguru Publications, 1986), p. 3.

The quotation from Descartes is from Cottingham, Stoothoff, and Murdoch (trans.), *The Philosophical Writings of Descartes*, Vol. II (Cambridge: Cambridge University Press, 1984), p. 374.

The quotations from G. E. Moore are from pp. 295, 296, and 300 of 'Proof of An External World', *Proceedings of the British Academy*, 25 (1939): 273–300.

Bertrand Russell's views on scepticism are explained in *The Problems of Philosophy* (London: Williams & Norgate, 1912).

Jonathan Vogel's spatial structure argument is from 'The Refutation of Scepticism', in Mattias Steup and Ernest Sosa (eds), *Contemporary Debates in Epistemology* (Oxford: Blackwell, 2005), pp. 72–84.

The most influential statement of semantic externalism is Saul Kripke's *Naming and Necessity* (Cambridge, MA: Harvard University Press, 1973). Putnam uses semantic externalism against scepticism in *Reason, Truth and History* (Cambridge: Cambridge University Press, 1981).The 'fledgling brain-in-a-vat' objection was raised by Anthony Brueckner, 1986, 'Brains in a Vat', *Journal of Philosophy*, 83(3): 148–67.

David Chalmers's essay 'The Matrix as Metaphysics' was published in *Philosophers Explore the Matrix* (New York: Oxford University Press, 2005), pp. 132–76.

Timothy Williamson draws the immune system analogy in 'Knowledge and Scepticism', in Frank Jackson and Michael Smith (eds), *The Oxford Handbook of Contemporary Philosophy* (New York: Oxford University Press, 2005), pp. 681–700.

Chapter 3: Rationalism and empiricism

The quotation from Paracelsus is from Jolande Jacobi and Norbert Guterman's *Paracelsus: Selected Writings* (Princeton: Princeton University Press, 1995), pp. 112–14.

Francesco Sizzi's claims about Jupiter are made on pp. 16–17 of his treatise *Understanding Astronomy, Optics and Physics* (Venice: 1611).
Montaigne advocates that we should not take sides between the earth-centred and the sun-centred models in his essay 'An Apology for Raymond Sébond', in *The Complete Essays of Montaigne*, trans. M. A. Screech (New York: Penguin, 1987), p. 642.
The quotations from Descartes are from volume II, pp. 4 and 12 of Cottingham, Stoothoff, and Murdoch's translation of *The Philosophical Writings of Descartes* (Cambridge: Cambridge University Press, 1984).
The quotations from Locke are from the following books, chapters, and sections of his *Essay Concerning Human Understanding*: 1.1.3, 4.7.9, 1.2.5, 4.3.18, 4.1.2, 4.2.14. P. H. Nidditch (ed.) (Oxford: Oxford University Press, 1979).

Chapter 4: The analysis of knowledge

Russell tells the clock story in *Human Knowledge: Its Scope and Limits* (London: Allen and Unwin, 1948), pp. 170–1.
Edmund Gettier's paper 'Is Justified True Belief Knowledge?' was published in *Analysis*, 23 (6) (1963): 121–3.
The no-false-belief rule was advocated by Michael Clark, 'Knowledge and Grounds: A Comment on Mr. Gettier's Paper', *Analysis*, 24 (2) (1963): 46–8. The objection that Clark's proposal would rule out clear cases of knowledge was pressed by John Turk Saunders and Naratan Champawat, 'Mr. Clark's Definition of "Knowledge"', *Analysis*, 25 (1) (1964): 8–9.
William Lycan argues for the rule that knowledge must not be essentially based on any false assumptions in 'On the Gettier Problem Problem', in Stephen Hetherington (ed.), *Epistemology Futures* (Oxford: Oxford University Press, 2006), pp. 148–68; against this proposal, see Ted Warfield, 'Knowledge from Falsehood', *Philosophical Perspectives*, 19 (1) (2005): 405–16, and Branden Fitelson, 'Strengthening the Case for Knowledge from Falsehood', *Analysis*, 70 (4) (2010): 666–9.
Alvin Goldman proposed the causal theory of knowledge in 'A Causal Theory of Knowing', *Journal of Philosophy*, 64 (12) (1967): 357–72. He formulates reliabilism in 'Discrimination and Perceptual Knowledge', *Journal of Philosophy*, 73 (20) (1976): 771–91.

The lottery objection to reliabilism is pressed by John Hawthorne in his book *Knowledge and Lotteries* (Oxford: Oxford University Press, 2004).

Linda Zagzebski's recipe for creating Gettier cases is in 'The Inescapability of Gettier Problems', *Philosophical Quarterly*, 44 (174) (1994): 65–73.

Matt Weiner argues that our use of 'know' is guided by inconsistent principles in 'The (Mostly Harmless) Inconsistency of Knowledge Ascriptions', *Philosophers' Imprint*, 9 (1) (2009): 1–25. Mark Kaplan argues that epistemology should focus on justification rather than knowledge in 'It's Not What You Know that Counts', *The Journal of Philosophy*, 82 (7) (1985): 350–63.

Timothy Williamson advocates abandoning the programme of analysis in Chapter 1 of *Knowledge and its Limits* (Oxford: Oxford University Press, 2000). The quotation is from p. 47.

The Ancient Gettier cases are drawn from Dharmottara's *Ascertainment of Knowledge* (*c*.770 CE). The 'distant fire' case follows Jonathan Stoltz's rendition in 'Gettier and Factivity in Indo-Tibetan Epistemology', *Philosophical Quarterly*, 57 (228) (2007): 394–415, and the 'desert mirage' case is drawn from Georges Dreyfus's presentation of it in *Recognizing Reality: Dharmakīrti's Philosophy and its Tibetan interpretations* (Albany: SUNY Press, 1997).

Gaṅgeśa's causal theory of knowledge is laid out in the perception chapter of his *Jewel of Reflection on the Truth about Epistemology*, Stephen Phillips and Ramanuja Tatacharya (trans.) (New York: American Institute of Buddhist Studies, 2004).

Chapter 5: Internalism and externalism

Nozick's tracking theory of knowledge is laid out in his *Philosophical Explanations* (Cambridge, MA: Harvard University Press, 1981).

Goldman acknowledges the problem of individuating belief-forming processes in 'Discrimination and Perceptual Knowledge', *The Journal of Philosophy*, 73 (20) (1976): 771–91; the problem is pressed against reliabilism by Earl Conee and Richard Feldman in 'The Generality Problem for Reliabilism', *Philosophical Studies*, 89.1 (1998): 1–29.

Juan Comesaña argues that, due to considerations of basing, the Generality Problem is everyone's problem, in 'A Well-Founded Solution to the Generality Problem', *Philosophical Studies*, 129.1 (2006): 27–47.

BonJour's Samantha example is from 'Externalist Theories of Empirical Knowledge', *Midwest Studies in Philosophy*, 5 (1) (1980): 53–74. BonJour lays out his objections to the idea of rescuing externalism by adding an internalist 'no-defeater' clause in *Epistemic Justification: Internalism vs. Externalism, Foundations vs. Virtues* (Malden, MA: Wiley-Blackwell, 2003).

The division between automatic and systematic thinking is explained in Jonathan Evans's *Thinking Twice: Two Minds in One Brain* (Oxford: Oxford University Press, 2010).

Chapter 6: Testimony

Locke's story about the King of Siam is from 4.15.5 of the *Essay Concerning Human Understanding*, and the checklist for determining the rational degree of confidence in testimony is from 4.15.4.

Dan Sperber discusses bees and people in 'An Evolutionary perspective on testimony and argumentation', *Philosophical Topics*, 29 (2001): 401–13. Ruth Millikan defends the view that even human testimonial knowledge transmission is 'cognition by proxy' in *Language, Thought and Other Biological Categories* (Cambridge: MIT Press, 1984).

The concept of 'Epistemic Vigilance' is explained in a paper of that title by Dan Sperber, Fabrice Clement, Christophe Heintz, Olivier Mascaro, Hugo Mercier, Gloria Origgi, and Deirdre Wilson, *Mind & Language*, 25 (4) (2010): 359–93.

The discussion of Gautama is based on Chapter 6 of Stephen Phillips's *Epistemology in Classical India* (Routledge 2012); the quotation (from an ancient commentary on Gautama) is from the *Nyāya-Sūtra* ed. A.M. Tarkatirtha, Taranatha Nyayatarkatirtha, and H. K. Tarkatirtha, 1936–45 (rpt.1985), as quoted by Phillips at p. 83 of *Epistemology in Classical India.*

The story of the lawyer is from *The Theaetetus of Plato*. Trans. M. J. Levett, ed. Myles Burnyeat (Indianapolis: Hackett Publishing Company, 1990), p. 338.

The quotation from Jennifer Lackey is from *Learning from Words*, (New York: Oxford University Press), p. 47.

The reliability of Wikipedia articles is reviewed in Jim Giles, 'Internet encyclopaedias go head to head', *Nature*, 438 (7070) (2005): 900–1.

Edward Craig's position on testimony is explained in his book *Knowledge and the State of Nature* (Oxford: Clarendon Press, 1990).

Chimpanzees' failure to distinguish between knowledgeable and ignorant informants is documented in Daniel Povinelli, Alyssa Rulf, and Donna Bierschwale's 'Absence of knowledge attribution and self-recognition in young chimpanzees', *Journal of Comparative Psychology* 108 (1) (1994): 74–80. Their ability to track what competitors know is described in Hare, B., Call, J., and Tomasello, M. (2001), 'Do chimpanzees know what conspecifics know?' *Animal Behaviour* 61 (1): 139–51.

Chapter 7: Shifting standards?

On the semantics of words like 'tall', see Peter Ludlow's 'Implicit Comparison Classes', *Linguistics and Philosophy*, 12 (1989): 519–33.

The Relevant Alternatives theory was advanced by Fred Dretske in 'Epistemic Operators', *Journal of Philosophy*, 64 (24) (1970): 1007–23.

Gail Stine introduces contextualism in 'Skepticism, Relevant Alternatives, and Deductive Closure', *Philosophical Studies*, 29 (4) (1976): 249–61. The quotation is from p. 254.

Stewart Cohen presents an internalist, evidence-centred version of contextualism in 'Contextualism, Skepticism, and the Structure of Reasons', *Philosophical Perspectives*, 13 (1999): 57–89; for externalist versions of contextualism, see David Lewis, 'Elusive Knowledge', *Australasian Journal of Philosophy*, 74 (1996): 549–67 and Keith DeRose, *The Case for Contextualism*, volume I (Oxford: Oxford University Press, 2009).

The point about respecting other speakers' contexts is made most clearly by Keith DeRose in 'Contextualism and Knowledge Attributions', *Philosophy and Phenomenological Research*, 52 (1992): 913–29. DeRose applies contextualism to the sceptical problem in 'Solving the Skeptical Problem', *The Philosophical Review*, 104 (1995): 1–52.

'Interest-relative invariantism' is a label coined by Jason Stanley; he defends the position in his book *Knowledge and Practical Interests* (Oxford: Oxford University Press, 2005).

Keith DeRose defends the idea that 'know(s)' could have its own kind of context-sensitivity in *The Case for Contextualism*, volume I

(Oxford: Oxford University Press, 2009); he also observes that IRI has trouble accounting for certain common patterns of intuition.

Timothy Williamson argues that contextualism misdiagnoses cases in which we are actually confused about the difference between knowing and knowing that we know, in 'Contextualism, subject-sensitive invariantism and knowledge of knowledge', *Philosophical Quarterly*, 55 (2005): 213–35. Patrick Rysiew argues that contextualists are mistaken about the pragmatics of knowledge attributions in 'The Context-Sensitivity of Knowledge Attributions', *Noûs* 35 (4) (2001): 477–514.

Chapter 8: Knowing about knowing

On primate inability to represent false belief even in competitive situations, see Juliane Kaminski, Josep Call, and Michael Tomasello, 'Chimpanzees know what others know, but not what they believe', *Cognition*, 109 (2) (2008): 224–34.

The hidden contents task is from Juergen Hogrefe, Heinz Wimmer and Josef Perner, 'Ignorance versus false belief: A developmental lag in attribution of epistemic states', *Child Development*, 57 (3) (1986): 567–82.

For a debate over the significance of developmental and comparative psychology to the knowledge-first programme, see Jennifer Nagel, 'Knowledge as a Mental State', and the replies to it, by Patrick Rysiew, Stephen Butterfill, and Johannes Roessler, in *Oxford Studies in Epistemology*, 4 (2012): 273–344.

Work on the early infant implicit recognition of false belief starts with Kristine Onishi and Renée Baillargeon, 'Do 15-month-old infants understand false beliefs?' *Science*, 308 (5719) (2005): 255–8. For discussion of some lingering questions about early infant mindreading, see Celia Heyes, 'False Belief in Infancy: A fresh look', *Developmental Science* (2014).

The short stories in Box 7 are from Rebecca Saxe, Susan Whitfield-Gabrieli, Jonathan Scholz, and Kevin Pelphrey, 'Brain Regions for Perceiving and Reasoning about Other People in School-Aged Children', *Child Development*, 80 (4) (2009): 1197–209. The role of the RTPJ in mindreading is explained in Rebecca Saxe and Nancy Kanwisher, 'People thinking about thinking people: the role of the temporo-parietal junction in the theory of mind', *Neuroimage*, 19 (2003): 1835–42.

Observations on the limits of nested mental state attribution are from Peter Kinderman, Robin Dunbar, and Richard P. Bentall's 'Theory-of-mind deficits and causal attributions', *British Journal of Psychology*, 89, no. 2 (1998): 191–204. On those with larger social circles doing better, see James Stiller and Robin Dunbar, 'Perspective-taking and memory capacity predict social network size', *Social Networks*, 29 (1) (2007): 93–104.

On the limits of the number of objects we can track at a time, see Zenon Pylyshyn and Ron Storm, 'Tracking multiple independent targets: Evidence for a parallel tracking mechanism', *Spatial Vision*, 3 (3) (1988): 179–97.

On egocentrism, see Susan Birch and Paul Bloom, 'Understanding children's and adults' limitations in mental state reasoning', *Trends in Cognitive Sciences*, 8 (6) (2004): 255–60. The piggy-bank experiment is from Daniela O'Neill, Janet Wilde Astington, and John Flavell's 'Young Children's Understanding of the Role that Sensory Experiences Play in Knowledge Acquisition', *Child Development*, 63 (1992): 474–90. The stock-trading game is from Colin Camerer, George Loewenstein, and Martin Weber's 'The curse of knowledge in economic settings: An experimental analysis', *The Journal of Political Economy*, 97 (5) (1989): 1232–54. The gambler experiment is from Jonathan Baron and John Hershey's 'Outcome bias in decision evaluation', *Journal of Personality and Social Psychology*, 54 (4) (1988): 569–79.

The impact of egocentrism on the intuitions motivating contextualism is explored in Jennifer Nagel's 'Knowledge Ascriptions and the Psychological Consequences of Thinking about Error', *Philosophical Quarterly*, 60 (239) (2010): 286–306.

The first major paper in experimental philosophy is Jonathan Weinberg, Shaun Nichols, and Stephen Stich's 'Normativity and Epistemic Intuitions', *Philosophical Topics*, 29 (2001): 429–60. Ernest Sosa criticizes the Weinberg paper in 'A Defense of the Use of Intuitions in Philosophy', in *Stich and his Critics*, Michael Bishop and Dominic Murphy, (eds), (Oxford: Blackwell, 2008), pp. 101–12.

Work documenting a failure to replicate the original Weinberg results includes Jennifer Nagel, Valerie San Juan, and Raymond Mar's 'Lay denial of knowledge for justified true beliefs', *Cognition*, 129 (2013): 652–6; this paper also includes the experimental findings concerning the story about Albert. John Turri also found no significant differences between North American and South Asian

Gettier case responses, in 'A conspicuous art: putting Gettier to the test', *Philosophers' Imprint*, 13 (10) (2013): 1–16.

Work on the universality of developmental stages in the acquisition of the concept of knowledge includes Henry Wellman, David Cross, and Julianne Watson's 'Meta-analysis of theory-of-mind development: the truth about false belief', *Child Development*, 72.3 (2001): 655–84, and David Liu, Henry M. Wellman, Twila Tardif, and Mark Sabbagh's, 'Theory of mind development in Chinese children: a meta-analysis of false-belief understanding across cultures and languages', *Developmental Psychology*, 44.2 (2008): 523–31. For the Chinese performance advantage, see Shali Wu and Boaz Keysar's 'The effect of culture on perspective taking', *Psychological Science*, 18 (7) (2007): 600–6.

Robert Cummins criticizes reliance on intuition in 'Reflection on Reflective Equilibrium', in Michael DePaul and William Ramsey (eds), *Rethinking Intuition: The Psychology of Intuition and Its Role in Philosophical Inquiry* (Oxford: Rowman & Littlefield, 1998), pp. 113–27; the quotation is from p. 124.

Further reading

Chapter 1: Introduction

The open-access archive Philpapers (<http://www.philpapers.org>) contains thousands of papers in epistemology, including most of those referenced here, organized into topic areas and searchable by keyword. In some cases the description of a paper links into a journal that is accessible only through a paywall (or from a computer with a subscription to the journal), but many of the papers are archived within Philpapers itself and are freely accessible.

There is a good discussion of group knowledge attributions in Alexander Bird's 'Social Knowing', *Philosophical Perspectives*, 24 (1) (2010): 23–56. For a clear survey of work on the relationship between individual and group judgement see Fabrizio Cariani's 'Judgment Aggregation', *Philosophy Compass*, 6 (2011): 22–32.

For a thorough examination of Protagoras's relativism, and Plato's response to it, see Myles Burnyeat's edition of Plato's *Theaetetus* (Indianapolis: Hackett, 1990) which includes a detailed and helpful introduction.

For a contemporary defence of relativism, see John MacFarlane's entry on relativism in *The Routledge Companion to the Philosophy of Language* (New York: Routledge, 2012).

Chapter 2: Scepticism

Charles Brittain's collection *On Academic Scepticism* contains the core texts and a useful introduction (Indianapolis: Hackett Publishing Company, 2006).

The lives and main ideas of ancient Greek sceptics are well covered in the *Stanford Encyclopedia of Philosophy* (online at <http://plato.stanford.edu/>).

One of the few positive contemporary defences of scepticism is Peter Unger's book, *Ignorance: A Case for Skepticism* (New York: Oxford University Press, 1975). Barry Stroud's *The Significance of Philosophical Scepticism* (New York: Oxford University Press, 1984) doesn't advocate scepticism, but Stroud takes the Dreaming Argument very seriously, and argues that there is still no fully satisfactory response to it.

Readers interested in Indian scepticism will enjoy Chapter two of Bimal Matital's book *Perception: An Essay on Classical Indian Theories of Knowledge* (Oxford: Oxford University Press, 1986).

Moore's position on the sceptical problem is now sometimes known as 'dogmatism'. For a revival of the Moorean way of looking at things, see James Pryor's 'The Skeptic and the Dogmatist', *Noûs*, 34 (4) (2000): 517–49.

Chapter 3: Rationalism and empiricism

Michael Matthew's collection *The Scientific Background to Modern Philosophy* (Indianapolis: Hackett, 1989) includes a wide range of interesting selections from scientists working in and just before the Early Modern period.

Gary Hatfield's *Stanford Encyclopedia of Philosophy* entry on Descartes (<http://plato.stanford.edu/entries/descartes/>) provides a good overview, as does Tom Sorrell's *Descartes: A Very Short Introduction* (Oxford: Oxford University Press, 2001). For more philosophical detail, readers can consult the essays in Janet Broughton and John Carreiro's *A Companion to Descartes* (Malden, MA: Blackwell, 2008). Readers interested in the life of Descartes will enjoy Stephen Gaukroger's *Descartes: An Intellectual Biography* (New York: Oxford University Press, 1995).

Descartes's *Meditations* were published with a series of objections by Descartes's contemporaries (including the prominent French theologian Antoine Arnauld and the prominent English philosopher Hobbes), together with Descartes's replies. These are freely available online, and in most full published editions of the *Meditations*. Descartes's correspondence with Elizabeth was not published in his day, but is now available in a translation by Lisa

Shapiro: *The Correspondence Between Princess Elizabeth of Bohemia and René Descartes* (Chicago: University of Chicago Press, 2007).

On Locke, after starting with the *Stanford Encyclopedia* entry by William Uzgalis (<http://plato.stanford.edu/entries/locke/>), readers can consult the *Cambridge Companion to Locke* (Cambridge: Cambridge University Press, 1994), and for more detail on the theory of knowledge, the *Cambridge Companion to Locke's Essay Concerning Human Understanding* (Cambridge: Cambridge University Press, 2007). Roger Woolhouse's *Locke: A Biography* (Cambridge: Cambridge University Press, 2007) is also recommended reading.

Chapter 4: The analysis of knowledge

Robert Shope's book *The Analysis of Knowing: A Decade of Research* (Princeton: Princeton University Press, 1983) is a great survey of all the early fights over Gettier's definition of knowledge.

More recent work is well covered in the *Stanford Encyclopedia of Philosophy* entry on the analysis of knowledge, by Jonathan Jenkins Ichikawa and Matthew Steup (<http://plato.stanford.edu/entries/knowledge-analysis/>).

Chapter 5: Internalism and externalism

Hilary Kornblith's collection *Epistemology: Internalism and Externalism* (Malden, MA: Wiley-Blackwell, 2001) is a good selection of the core early writings on either side of the internalism–externalism controversy.

The classic statement of 20th century internalism is Roderick Chisholm's *Theory of Knowledge* (Englewood Cliffs, NJ: Prentice-Hall, 1966).

Externalism is defended in Alvin Goldman's *Epistemology and Cognition* (Cambridge, MA: Harvard University Press, 1986) and in Timothy Williamson's *Knowledge and its Limits* (Oxford: Oxford University Press, 2000).

For a clear summary of the internalism–externalism controversy, see James Pryor's 2001 paper 'Highlights of Recent Epistemology', *British Journal for the Philosophy of Science*, 52: 95–124.

Chapter 6: Testimony

For a concise introduction to reductionism and non-reductionism, see Jennifer Lackey's 'Knowing from Testimony', *Philosophy Compass*, 1:5 (2006): 432–48. Jennifer Lackey and Ernest Sosa's edited collection *The Epistemology of Testimony* contains essays representing a broad spectrum of philosophical perspectives on testimony (Oxford: Oxford University Press, 2006).

For more on the classical Indian line on testimony, see Stephen Phillips's *Stanford Encyclopedia* entry on Classical Indian Epistemology: <http://plato.stanford.edu/entries/epistemology-india/>.

Chapter 7: Shifting standards?

Patrick Rysiew's *Stanford Encyclopedia of Philosophy* entry on contextualism is an excellent overview of the position: <http://plato.stanford.edu/entries/contextualism-epistemology/>.

For a collection of influential essays for and against contextualism, see *Contextualism in Philosophy: Knowledge, Meaning and Truth*, Gerhard Preyer and Georg Peter (eds) (New York: Oxford University Press, 2005).

Chapter 8: Knowing about knowing

For an overview of empirical work on mindreading, see Ian Apperly's *Mindreaders: The Cognitive Basis of 'Theory of Mind'* (Hove: Psychology Press, 2011).

For various perspectives on experimental philosophy, see *Current Controversies in Experimental Philosophy*, Edouard Machery and Elizabeth O'Neill (eds) (New York: Routledge, 2014).

There are interesting discussions of the role of intuitions in philosophy in Hilary Kornblith's *Knowledge and its Place in Nature* (Oxford: Oxford University Press, 2005), in Timothy Williamson's *The Philosophy of Philosophy* (Oxford: Oxford University Press, 2007), and in Tamar Szabó Gendler's *Intuition, Imagination, and Philosophical Methodology* (Oxford University Press, 2010).

“牛津通识读本”已出书目

古典哲学的趣味
人生的意义
文学理论入门
大众经济学
历史之源
设计，无处不在
生活中的心理学
政治的历史与边界
哲学的思与惑
资本主义
美国总统制
海德格尔
我们时代的伦理学
卡夫卡是谁
考古学的过去与未来
天文学简史
社会学的意识
康德
尼采
亚里士多德的世界
西方艺术新论
全球化面面观
简明逻辑学
法哲学：价值与事实
政治哲学与幸福根基
选择理论
后殖民主义与世界格局
福柯
缤纷的语言学
达达和超现实主义
佛学概论
维特根斯坦与哲学
科学哲学
印度哲学祛魅
克尔凯郭尔
科学革命
广告
数学
叔本华
笛卡尔
基督教神学
犹太人与犹太教
现代日本
罗兰·巴特
马基雅维里
全球经济史
进化
性存在
量子理论
牛顿新传
国际移民
哈贝马斯
医学伦理
黑格尔
地球
记忆
法律
中国文学
托克维尔
休谟
分子
法国大革命
民族主义
科幻作品
罗素
美国政党与选举
美国最高法院
纪录片
大萧条与罗斯福新政
领导力
无神论
罗马共和国
美国国会
民主
英格兰文学
现代主义
网络
自闭症
德里达
浪漫主义
批判理论

德国文学
戏剧
腐败
医事法
癌症
植物
法语文学
微观经济学
湖泊
拜占庭
儿童心理学
时装
现代拉丁美洲文学
卢梭
隐私
电影音乐
抑郁症
传染病
希腊化时代
知识
电影
俄罗斯文学
古典文学
大数据
洛克
幸福
免疫系统
银行学
景观设计学